UNITEXT for Physics

T0172048

UNITEXT for Physics series, formerly UNITEXT Collana di Fisica e Astronomia, publishes textbooks and monographs in Physics and Astronomy, mainly in English language, characterized of a didactic style and comprehensiveness. The books published in UNITEXT for Physics series are addressed to graduate and advanced graduate students, but also to scientists and researchers as important resources for their education, knowledge and teaching.

More information about this series at http://www.springer.com/series/13351

Leonardo Angelini

Solved Problems in Quantum Mechanics

 Springer

Leonardo Angelini
Bari University
Bari, Italy

ISSN 2198-7882   ISSN 2198-7890   (electronic)
UNITEXT for Physics
ISBN 978-3-030-18406-3   ISBN 978-3-030-18404-9   (eBook)
https://doi.org/10.1007/978-3-030-18404-9

This Springer imprint is published by the registered company Springer Nature Switzerland AG
The registered company address is: Gewerbestrasse 11, 6330 Cham, Switzerland

Preface

This book is essentially devoted to students who wish to prepare for written examinations in a Quantum Mechanics course. As a consequence, this collection can also be very useful for teachers who need to propose problems to their students, both in class and in examinations. Like many other books of Quantum Mechanics Problems, one should not expect a particular novel effort. The aim is to present problems that, in addition to exploring the student's understanding of the subject and their ability to apply it concretely, are solvable in a limited time. This purpose is unlikely to be combined with a search for originality.

Problems will therefore be found that are also present in other books from the Russian classics [1, 2], and, therefore, in the collection, extracted from them, cared for by Ter Haar [3, 4]. Among other books of exercises that have been consulted are the Italian Passatore [5] and that most recently published by Yung-Kuo Lim [6], which collects the work of 19 Chinese physicists. The two volumes by Flügge [7] lie between a manual and a problem book, providing useful tips, though the presented problems are often too complex in relation to the purpose of this collection.

Many interesting problems are also found in Quantum Mechanics manuals. In this case, the list could be very long. I will only mention those who have devoted more space to problems: the classical manuals of Merzbacher [8] and Gasiorowicz [9], the volume devoted to Quantum Mechanics in the Theoretical Physics course by Landau and Lifchitz [10], the two volumes by Messiah [11] and the most recent works by Shankar [12], Gottfried-Yan [13], and Sakurai-Napolitano [14]. One particular quote is due to Nardulli's Italian text [15], both because of the abundance of problems it contains with or without solution, and the fact that many problems presented here have been proposed over the years to students of his course.

The category of *problems that can be resolved in a reasonable time* is not the only criterion for our choice. No problem has been included that requires knowledge of mathematical methods that are sometimes absent from standard courses, such as, for example, Fuchsian differential equations. When necessary, complementary mathematical formulas have been included in the appendix. The most important characteristic of this book is that the solutions of many problems are presented with some detail, eliminating only the simplest steps. This will certainly prove useful to the students. Like in any other book, problems have been grouped into chapters. In many cases, the inclusion of a particular problem in a particular chapter can be considered arbitrary: many exam problems pose cross-cutting issues across the entire program. The obvious choice was to take into account the most distinctive questions.

For a time, this collection was entrusted to the network and used by teachers and students. It is thanks to some of them that many of the errors initially present have been eliminated. I thank Prof. Stefano Forte for encouraging me to publish it in print after completing certain parts and reviewing the structure. One last great thanks goes to my wife; the commitment needed to draft this text also resulted in a great deal of family burdens falling on her.

Finally, I apologize to the readers for the errors that surely escaped me; every indication and suggestion is certainly welcome.

Bari, Italy   Leonardo Angelini
December 2018

Contents

List of Figures

Chapter 1
Operators and Wave Functions

1.1 Spectrum of Compatible Variables

Given three variables A, B, C, demonstrate that if $[A, B]=[A, C] = 0$, but $[B, C] \neq 0$, the spectrum of A is degenerate.

Solution

Suppose that all of the eigenvalues of A are not degenerate, so that, for each eigenvalue a of A, there is only one ket $|\psi_a\rangle$ such that

$$A|\psi_a\rangle = a|\psi_a\rangle.$$

If this were true, each ket $|\psi_a\rangle$ must also be eigenstate of B and C that are compatible with A. As a consequence, we can also label the ket $|\psi_a\rangle$ with the eigenvalues of B e C:

$$A|\psi_{a,b,c}\rangle = a|\psi_{a,b,c}\rangle$$
$$B|\psi_{a,b,c}\rangle = b|\psi_{a,b,c}\rangle$$
$$C|\psi_{a,b,c}\rangle = c|\psi_{a,b,c}\rangle$$

where, obviously, once a is fixed, b e c must be unique. For each generic state $|\psi\rangle$, it results that

$$[B, C]|\psi\rangle = (BC - CB) \sum_a |\psi_{a,b,c}\rangle = \sum_a (bc - cb)|\psi_{a,b,c}\rangle = 0.$$

This contradicts our initial supposition that $[B, C] \neq 0$.

© Springer Nature Switzerland AG 2019
L. Angelini, *Solved Problems in Quantum Mechanics*, UNITEXT for Physics,
https://doi.org/10.1007/978-3-030-18404-9_1

1.2 Constants of Motion

Show that, if F e G are two constants of motion for a quantum system, this is also true for $[F, G]$.

Solution

If F and G are two constants of motion, then, from the Heisemberg equation,

$$\frac{\partial F}{\partial t} = \frac{i}{\hbar}[F, \mathcal{H}] \quad \text{and} \quad \frac{\partial G}{\partial t} = \frac{i}{\hbar}[G, \mathcal{H}],$$

where $\mathcal{H}$ is the system Hamiltonian. It turns out that

$$\frac{d}{dt}[F, G] = \frac{\partial[F, G]}{\partial t} - \frac{i}{\hbar}[[F, G], \mathcal{H}] =$$

$$= \frac{\partial F}{\partial t}G + F\frac{\partial G}{\partial t} - \frac{\partial G}{\partial t}F - G\frac{\partial F}{\partial t} - \frac{i}{\hbar}[FG - GF, \mathcal{H}] =$$

$$= \frac{i}{\hbar}[F\mathcal{H}G - \mathcal{H}FG + FG\mathcal{H} - F\mathcal{H}G - G\mathcal{H}F + \mathcal{H}GF - GF\mathcal{H} + G\mathcal{H}F - $$

$$- FG\mathcal{H} + GF\mathcal{H} + \mathcal{H}FG - \mathcal{H}GF] = 0.$$

Hence, $[F, G]$ is a constant of motion.

1.3 Number Operator

Let an operator a be given that satisfies the following relationships:

$$aa^+ + a^+a = 1,$$

$$a^2 = (a^+)^2 = 0.$$

(a) Can operator a be hermitian?
(b) Prove that the only possible eigenvalues for operator $N = a^+a$ are 0 and 1.

Solution

(a) Suppose that a is hermitian: $a = a^+$. We obtain

$$aa^+ + a^+a = 2(a^+)^2 = 0,$$

which contradicts the initial statement.
(b) $N^2 = a^+aa^+a = a^+(1 - a^+a)a = a^+a - (a^+)^2a^2 = a^+a = N.$
It is well known that, if an operator satisfies an algebraic equation, this is also

satisfied by its eigenvalues. Indeed, calling $|\lambda\rangle$ the generic eigenket of N corresponding to the eigenvalue λ, we can write

$$(N^2 - N)|\lambda\rangle = (\lambda^2 - \lambda)|\lambda\rangle = 0 \Rightarrow \lambda = 0, 1$$

1.4 Momentum Expectation Value

Given a particle of mass m in a potential $V(\mathbf{r})$, system described by the Hamiltonian

$$H = T + V = \frac{p^2}{2m} + V(\mathbf{r}),$$

demonstrate the relationship

$$\mathbf{p} = -i\frac{m}{\hbar}[\mathbf{r}, \mathcal{H}].$$

Use this relationship to show that, in a stationary state,

$$\langle \mathbf{p} \rangle = 0.$$

Solution

Calling r_i and p_i ($i = 1, 2, 3$) the position and momentum components, we have

$$\begin{aligned}
[r_i, \mathcal{H}] = [r_i, T] &= \frac{1}{2m}(r_i p_i^2 - p_i^2 r_i) = \\
&= \frac{1}{2m}(r_i p_i^2 - p_i^2 r_i - p_i r_i p_i + p_i r_i p_i) = \\
&= \frac{1}{2m}([r_i, p_i]p_i + p_i[r_i, p_i]) = \\
&= \frac{i\hbar p_i}{m},
\end{aligned}$$

as conjectured. Calling $|\psi_E\rangle$ the eigenstate of $\mathcal{H}$ corresponding to an eigenvalue E, the expectation value of each momentum component is

$$\begin{aligned}
\langle p_i \rangle = \langle \psi_E | p_i | \psi_E \rangle &= -i\frac{m}{\hbar}\langle \psi_E |[r_i, \mathcal{H}]|\psi_E \rangle = \\
&= -i\frac{m}{\hbar}\Big[\langle \psi_E | r_i E |\psi_E \rangle - \langle \psi_E | E r_i |\psi_E \rangle\Big] = 0,
\end{aligned}$$

provided $\langle r_i \rangle$ is a well-defined quantity. Indeed, this result is invalid for improper eigenvectors: this is the case of free particles, when you consider $|\psi_E\rangle$ as a simultaneous eigenstate of $\mathcal{H}$ and $\mathbf{p}$ or, generally, an eigenstate of the continuous part of the spectrum.

1.5 Wave Function and the Hamiltonian

A particle is in a state described by the following wave function:

$$\psi(\mathbf{r}) = A \sin\left(\frac{\mathbf{p} \cdot \mathbf{r}}{\hbar}\right).$$

(a) Is it a free particle?
(b) What can we say about the value of momentum and energy in this state?

Solution

(a) The wave function is representative of the dynamical state of a system. To decide whether the particle is free, we need to know the Hamiltonian.
(b) We can write this wave function as

$$\psi(\mathbf{r}) = \frac{A}{2i}\left(e^{i\frac{\mathbf{p} \cdot \mathbf{r}}{\hbar}} - e^{-i\frac{\mathbf{p} \cdot \mathbf{r}}{\hbar}}\right).$$

Clearly, it represents the superposition of two momentum eigenstates with eigen-values $+\mathbf{p}$ and $-\mathbf{p}$. As the coefficients of the linear superposition have equal magnitude, the momentum expectation value is zero. Without the knowledge of the Hamiltonian, it is impossible to say anything about the energy.

1.6 What Does a Wave Function Tell Us?

A particle constrained to move in one dimension is described at a certain instant by the wave function

$$\psi(x) = A \cos kx.$$

Can we infer that:

(a) it describes a state with defined momentum?
(b) it describes a free particle state?

Solution

(a) The wave function can be written as

$$\psi(x) = \frac{A}{2}\left(e^{ikx} + e^{-ikx}\right).$$

It is the linear superposition of two momentum eigenstates with momentum $p = \hbar k$ and $p = -\hbar k$. As they have equal amplitudes, they are equiprobable. So, the answer to the question is no. The kinetic energy $E = \frac{p^2}{2m}$ is defined instead.

(b) The question has no answer. The wave function *could* be the eigenfunction of a potential free Hamiltonian. Nevertheless, it should be remembered that the wave function specifies the state of a system, not its dynamic. Instead, the dynamics is specified by the Hamiltonian, which, in this case, is unknown.

1.7   Spectrum of a Hamiltonian

Consider a physical system described by the Hamiltonian

$$\mathcal{H} = \frac{p^2}{2m} + \frac{\alpha}{2}(pq + qp) + \beta q^2, \qquad [q, p] = i\hbar.$$

Find the α and β values for which $\mathcal{H}$ is bounded from below and, if this is the case, find its eigenvalues and eigenvectors.

Solution

The Hamiltonian can be rewritten as:

$$\mathcal{H} = \frac{1}{2m}[p^2 + \alpha m(pq + qp) + m^2\alpha^2 q^2 - m^2\alpha^2 q^2 + \beta q^2] =$$

$$= \frac{1}{2m}(p + m\alpha q)^2 + \left(\beta - \frac{m\alpha^2}{2}\right)q^2 =$$

$$= \frac{1}{2m}p'^2 + \left(\beta - \frac{m\alpha^2}{2}\right)q^2,$$

where

$$p' = p + m\alpha q.$$

Note now that:

- $[q, p'] = [q, p + m\alpha q] = [q, p] = i\hbar;$
- p' is hermitian, being a linear combination of two hermitian operators, provided α is real.

(Note that these properties also apply in the case of $p' = p + f(q)$, with $f(q)$ being a real function of q.)

Impose that $\mathcal{H}$ is bounded from below:

$$\langle\psi|\mathcal{H}|\psi\rangle = \frac{1}{2m}\langle p'\psi|p'\psi\rangle + (\beta - \frac{m\alpha^2}{2})\langle q\psi|q\psi\rangle > -\infty.$$

The first term being positive or zero, this condition is verified for every $|\psi\rangle$, provided

$$\beta > \frac{m\alpha^2}{2}.$$

If this is true, we can replay the harmonic oscillator procedure for this Hamiltonian, obtaining the same eigenvalues and eigenstates. In this case, the frequency is

$$\omega = \sqrt{\frac{2\beta}{m} - \alpha^2}.$$

1.8 Velocity Operator for a Charged Particle

Given a a charged particle in a magnetic field, find the commutation relations between the operators corresponding to the velocity components.

Solution

Remember that the Hamiltonian of a particle having charge q in an electromagnetic field is

$$\mathcal{H} = \frac{1}{2m} \left(\mathbf{P} - \frac{q}{c} \mathbf{A} \right)^2 + q \phi,$$

where $\mathbf{A}$ and ϕ are the magnetic and electric potential giving rise to the electromagnetic field: $\mathbf{B} = \nabla \times \mathbf{A}$, $\mathbf{E} = -\frac{1}{c} \frac{\partial \mathbf{A}}{\partial t} - \nabla \phi$.

$\mathbf{P}$ is the canonical momentum, i.e., the momentum conjugate to the coordinate $\mathbf{r}$ and corresponding, in Quantum Mechanics, to the operator $-i\hbar\nabla$ (coordinate representation). The velocity, instead, is obtained from

$$\mathbf{v} = \nabla_{\mathbf{P}} \mathcal{H} = \frac{1}{m} \left(\mathbf{P} - \frac{q}{c} \mathbf{A} \right).$$

Thus, in the coordinate representation, the velocity components operators are

$$v_i = \frac{1}{m} \left(P_i - \frac{q}{c} A_i \right) = \frac{1}{m} \left(-i\hbar \frac{\partial}{\partial x_i} - \frac{q}{c} A_i \right),$$

where the components of $\mathbf{A}$ are not operators. Thus, the desired commutators are

$$[v_i, v_j] \psi(\mathbf{r}) = \frac{1}{m^2} \left[P_i - \frac{q}{c} A_i, P_j - \frac{q}{c} A_j \right] \psi(\mathbf{r}) =$$

$$= \frac{q}{mc^2} \left\{ [P_j, A_i] - [P_i, A_j] \right\} \psi(\mathbf{r}) =$$

$$= \frac{i\hbar q}{mc^2} \left(A_i \frac{\partial \psi(\mathbf{r})}{\partial x_j} - \frac{\partial A_j \psi(\mathbf{r})}{\partial x_i} + A_j \frac{\partial \psi(\mathbf{r})}{\partial x_i} - \frac{\partial A_i \psi(\mathbf{r})}{\partial x_j} \right) =$$

$$= \frac{i\hbar q}{mc^2} \left(\frac{\partial A_j}{\partial x_i} - \frac{\partial A_i}{\partial x_j} \right) \psi(\mathbf{r}) =$$

$$= \frac{i\hbar q}{mc^2} \varepsilon_{ijk} B_k \psi(\mathbf{r}),$$

where ε_{ijk} is the Levi-Civita symbol.

1.9 Power-Law Potentials and Virial Theorem

A one-dimensional system is described by the Hamiltonian

$$\mathcal{H} = \frac{p^2}{2m} + \lambda q^n.$$

Given an eigenstate $|\psi\rangle$ of this Hamiltonian, prove that

$$\langle \mathcal{T} \rangle = \langle \psi | \mathcal{T} | \psi \rangle = \frac{n}{2} \langle \psi | \mathcal{V} | \psi \rangle = \frac{n}{2} \langle \mathcal{V} \rangle,$$

where $\mathcal{T} = p^2/2m$ e $\mathcal{V}$ is the potential energy $\mathcal{V} = \lambda q^n$.

Solution

Note that

$$
\begin{aligned}
[q, \mathcal{H}] = [q, \mathcal{T}] &= \frac{1}{2m} (qp^2 - p^2 q) = \\
&= \frac{1}{2m} (qp^2 - p^2 q - pqp + pqp) = \\
&= \frac{1}{2m} ([q, p]p + p[q, p]) = \\
&= \frac{i\hbar p}{m}.
\end{aligned}
$$

Using the coordinate representation, it is easy to verify that

$$q[p, \mathcal{H}] = q[p, \mathcal{V}] = \frac{\hbar}{i} \lambda n q^n = \frac{\hbar n}{i} \mathcal{V}.$$

So,

$$
\begin{aligned}
\langle \psi | \mathcal{V} | \psi \rangle &= -\frac{1}{i n \hbar} \langle \psi | q[p, \mathcal{H}] | \psi \rangle = \\
&= -\frac{1}{i n \hbar} \langle \psi | qp\mathcal{H} - q\mathcal{H}p | \psi \rangle = \\
&= -\frac{1}{i n \hbar} \langle \psi | qp\mathcal{H} - [q, \mathcal{H}]p - \mathcal{H}qp | \psi \rangle = \\
&= \frac{1}{i n \hbar} \langle \psi | [q, \mathcal{H}]p | \psi \rangle.
\end{aligned}
$$

Using the previous result, we obtain the desired relationship

$$\langle \psi | \mathcal{V} | \psi \rangle = \frac{1}{i n \hbar} \frac{i\hbar}{m} \langle \psi | p^2 | \psi \rangle = \frac{2}{n} \langle \psi | \mathcal{T} | \psi \rangle.$$

1.10   Coulomb Potential and Virial Theorem

(a) Using the Schrödinger equation, prove that, for every physical quantity Ω of a quantum system, the Ehrenfest theorem holds:

$$\frac{d\langle\Omega\rangle}{dt} = \frac{1}{\imath\hbar}\langle[\Omega,\mathcal{H}]\rangle + \left\langle\frac{\partial\Omega}{\partial t}\right\rangle.$$

(b) Apply this result to the operator $\mathbf{r}\cdot\mathbf{p}$ and prove the Virial theorem for the Coulomb potential, which relates the expectation values in a stationary state of the kinetic energy $\mathcal{T}$ and of the potential energy $\mathcal{V}$:

$$\langle\mathcal{T}\rangle = -\frac{1}{2}\langle\mathcal{V}\rangle.$$

Solution

(a) Call $|\psi(t)\rangle$ the state vector of the physical system in the instant t. The expectation value of Ω is

$$\langle\Omega\rangle = \langle\psi(t)|\Omega|\psi(t)\rangle,$$

and, from the Schrödinger equation

$$\imath\hbar\frac{\partial|\psi(t)\rangle}{\partial t} = \mathcal{H}|\psi(t)\rangle,$$

we get

$$\frac{d\langle\Omega\rangle}{dt} = \frac{\partial\langle\psi(t)|}{\partial t}\Omega|\psi(t)\rangle + \langle\psi(t)|\frac{\partial\Omega}{\partial t}|\psi(t)\rangle + \langle\psi(t)|\Omega\frac{\partial|\psi(t)\rangle}{\partial t} =$$

$$= \frac{1}{\imath\hbar}\langle\psi(t)|(\Omega\mathcal{H} - \mathcal{H}\Omega)|\psi(t)\rangle + \left\langle\frac{\partial\Omega}{\partial t}\right\rangle = \frac{1}{\imath\hbar}\langle[\Omega,\mathcal{H}]\rangle + \left\langle\frac{\partial\Omega}{\partial t}\right\rangle,$$

where we have taken into account the fact that, in the Schrödinger picture, time dependence of operators can only be explicit.

(c) As the system is in a stationary state and $\mathbf{r}\cdot\mathbf{p}$ does not depend on time,

$$\frac{d\langle\mathbf{r}\cdot\mathbf{p}\rangle}{dt} = \frac{\partial\langle\mathbf{r}\cdot\mathbf{p}\rangle}{\partial t} = 0.$$

Applying the Ehrenfest theorem, we obtain:

$$0 = \langle[\mathbf{r}\cdot\mathbf{p},\mathcal{H}]\rangle = \langle[\mathbf{r}\cdot\mathbf{p},\mathcal{T}]\rangle + \langle[\mathbf{r}\cdot\mathbf{p},\mathcal{V}]\rangle = \langle[\mathbf{r},\mathcal{T}]\cdot\mathbf{p}\rangle + \langle\mathbf{r}\cdot[\mathbf{p},\mathcal{V}]\rangle,$$

To calculate these two expectation values, we note that, from $[r_i, p_i] = \imath\hbar$, we get

$$[r_i, p_i^2] = 2\iota\hbar p_i \;\Rightarrow\; \langle[\mathbf{r}, \mathcal{T}]\cdot\mathbf{p}\rangle = 2\iota\hbar\langle\mathcal{T}\rangle,$$

whereas we easily find that

$$[\nabla, \frac{1}{r}] = -\frac{\mathbf{r}}{r^3} \;\Rightarrow\; \langle\mathbf{r}\cdot[\mathbf{p}, \mathcal{V}]\rangle = \iota\hbar e^2\langle\frac{1}{r}\rangle = \iota\hbar\langle\mathcal{V}\rangle.$$

By replacing these two relations in the previous one, the desired result is obtained.

1.11 Virial Theorem for a Generic Potential

(a) Using the Schrödinger equation, prove that, for every quantity Ω of a given physical system, the Ehrenfest theorem holds:

$$\frac{d\langle\Omega\rangle}{dt} = \frac{1}{\iota\hbar}\langle[\Omega, \mathcal{H}]\rangle + \left\langle\frac{\partial\Omega}{\partial t}\right\rangle.$$

(b) Consider a system with N degrees of freedom and apply the previous result in the case of the operator

$$Q = \sum_{i=1}^{N} r_i\, p_i,$$

with the purpose of demonstrating the Virial theorem relating the expectation values, in a stationary state, of the kinetic energy $\mathcal{T}$ and of the potential energy $\mathcal{V}$ (not dependent on time):

$$\langle\mathcal{T}\rangle = \frac{1}{2}\sum_{i=1}^{N}\left\langle q_i\frac{\partial\mathcal{V}}{\partial q_i}\right\rangle.$$

(c) Apply the previous result to a one-dimensional harmonic oscillator.

Solution

(a) For the solution of this point, we refer to Problem 1.10.
(b) Denoting the set of position coordinates with q and the set of conjugate momenta with p, we have $\mathcal{H}(q, p) = \mathcal{T}(p) + \mathcal{V}(q)$, and

$$[Q, \mathcal{H}] = \sum_{i=1}^{N}[q_i\, p_i, \mathcal{H}] = \sum_{i=1}^{N}[q_i, \mathcal{T}]p_i + \sum_{i=1}^{N}q_i[p_i, \mathcal{V}]. \tag{1.1}$$

We calculate the commutators on the right side separately:

$$[q_i, \mathcal{T}] = \frac{1}{2m_i} [q_i, p_i^2] = \frac{1}{2m_i} (q_i p_i^2 - p_i^2 q_i) = \frac{1}{2m_i} (p_i^2 q_i + 2i\hbar p_i - p_i^2 q_i) = \frac{i\hbar p_i}{m_i},$$

while, from the power expansion in the variables q_i of the potential $\mathcal{V} = \sum_n c_n q_i^n$, we obtain

$$[p_i, \mathcal{V}] = \sum_n c_n [p_i, q_i^n] = \sum_n c_n [p_i q_i^n - q_i^n p_i] = \sum_n c_n [q_i^n p_i - i\hbar n q_i^{n-1} - q_i^n p_i] =$$

$$= -i\hbar \sum_n c_n n q_i^{n-1} = -i\hbar \frac{\partial \mathcal{V}}{\partial q_i}.$$

By replacing these two relations in Equation (1.1) we get

$$[Q, \mathcal{H}] = 2i\hbar \mathcal{T} - i\hbar \sum_{i=1}^{N} q_i \frac{\partial \mathcal{V}}{\partial q_i}. \tag{1.2}$$

Applying the Ehrenfest theorem to the observable Q, under the hypothesis that the system is in a stationary state, we obtain

$$\frac{d\langle Q \rangle}{dt} = \frac{1}{i\hbar} \langle [Q, \mathcal{H}] \rangle + \left\langle \frac{\partial Q}{\partial t} \right\rangle.$$

Noting that neither Q, nor the probability distribution in a stationary state depends on time, we get

$$\langle [Q, \mathcal{H}] \rangle = 0 \quad \Rightarrow \quad 2\langle \mathcal{T} \rangle = \sum_{i=1}^{N} \left\langle q_i \frac{\partial \mathcal{V}}{\partial q_i} \right\rangle.$$

(c) The potential energy of a harmonic oscillator is

$$\mathcal{V}(x) = \frac{1}{2} m\omega^2 x^2.$$

Then, the kinetic energy expectation value is given by

$$\langle \mathcal{T} \rangle = \frac{x}{2} \left\langle \frac{d\mathcal{V}}{dx} \right\rangle = \langle \frac{1}{2} m\omega^2 x^2 \rangle = \langle \mathcal{V} \rangle.$$

1.12   Feynman-Hellmann Theorem

Given a physical system, denote its Hamiltonian with $\mathcal{H}$ having eigenvalues E and normalized eigenstates $|E\rangle$ so that

$$\mathcal{H}|E\rangle = E|E\rangle.$$

Assume that this Hamiltonian depends on a parameter λ, $\mathcal{H} = \mathcal{H}(\lambda)$. As a consequence, its eigenvalues also depend on λ, $E = E(\lambda)$.

Demonstrate that the following relationship holds:

$$\left\langle \frac{\partial \mathcal{H}(\lambda)}{\partial \lambda} \right\rangle = \frac{\partial E(\lambda)}{\partial \lambda}. \tag{1.3}$$

Solution

As E is the eigenvalue corresponding to $|E\rangle$, it results that

$$E(\lambda) = \langle E|\mathcal{H}(\lambda)|E\rangle.$$

It follows that

$$\frac{\partial E(\lambda)}{\partial \lambda} = \frac{\partial}{\partial \lambda} \langle E|\mathcal{H}(\lambda)|E\rangle =$$

$$= \frac{\partial \langle E|}{\partial \lambda} \mathcal{H}(\lambda)|E\rangle + \langle E|\frac{\partial \mathcal{H}(\lambda)}{\partial \lambda}|E\rangle + \langle E|\mathcal{H}(\lambda)\frac{\partial |E\rangle}{\partial \lambda} =$$

$$= \left\langle \frac{\partial \mathcal{H}(\lambda)}{\partial \lambda} \right\rangle + E\left(\frac{\partial \langle E|}{\partial \lambda}|E\rangle + \langle E|\frac{\partial |E\rangle}{\partial \lambda} \right) =$$

$$= \left\langle \frac{\partial \mathcal{H}(\lambda)}{\partial \lambda} \right\rangle + E\left(\frac{\partial}{\partial \lambda} \langle E|E\rangle \right) =$$

$$= \left\langle \frac{\partial \mathcal{H}(\lambda)}{\partial \lambda} \right\rangle,$$

as we wanted to prove.

Chapter 2
One-Dimensional Systems

2.1 Free Particles and Parity

For a one-dimensional free particle does the set of observables composed of the Hamiltonian and Parity constitute a complete set?

Solution

To every value of the energy of a free particle in a one-dimensional world, they correspond two linearly independent eigenstates that, in the **X**-representation, are given by the eigenfunctions:

$$\psi_p(x) = \frac{1}{\sqrt{2\pi\hbar}} e^{i\frac{px}{\hbar}} \quad \text{and} \quad \psi_{-p}(x) = \frac{1}{\sqrt{2\pi\hbar}} e^{-i\frac{px}{\hbar}}.$$

Any linear superposition of them is also an eigenstate of the Hamiltonian. As the Hamiltonian is given by the kinetic energy operator, which is proportional to the second x-derivative, it commutes with the Parity operator. The linear superposition of $\psi_p(x)$ e $\psi_{-p}(x)$ with definite parity is, neglecting normalization, $\cos(px/\hbar)$ with parity eigenvalue $P = +1$ and $\sin(px/\hbar)$ with parity $P = -1$. The eigenstates common to each pair of Hamiltonian and Parity eigenvalues are completely determined, so these two operators are a complete set of commuting observables for this physical system.

2.2 Potential Step

Consider a particle of mass m incident from the left on the potential (Fig. 2.1)

$$V(x) = \begin{cases} 0, & \text{for } x < 0; \\ V_0, & \text{for } x > 0. \end{cases}$$

© Springer Nature Switzerland AG 2019
L. Angelini, *Solved Problems in Quantum Mechanics*, UNITEXT for Physics,
https://doi.org/10.1007/978-3-030-18404-9_2

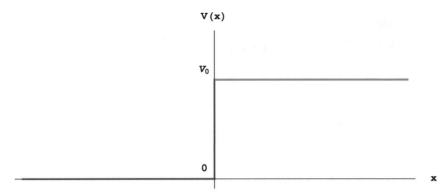

Fig. 2.1 Potential step

Study its behavior for energy eigenvalues lesser and greater than V_0, determining the energy spectrum characteristics.

Solution

The energy eigenvalues E are positive, because, as it is well known, they must be greater than the potential minimum. Let's calculate the Schrödinger equation solutions in each of the two regions where the potential stays constant.

Region I: $x < 0$

In this region, the Schrödinger equation can be written as:

$$-\frac{\hbar^2}{2m} \frac{d^2 \psi(x)}{dx^2} = E \psi(x),$$

that is,

$$\frac{d^2 \psi(x)}{dx^2} + k^2 \psi(x),$$

where

$$k = \sqrt{\frac{2mE}{\hbar^2}} > 0.$$

The most general solution is given by the linear superposition

$$\psi(x) = e^{ikx} + Re^{-ikx}, \tag{2.1}$$

where we put the coefficient of the incident plane wave propagating to the right as being equal to 1, while R is the coefficient of the reflected wave. The physical meaning of this assumption is obtained calculating the probability density current from (A.11):

$$j(x) = \frac{\hbar}{m} \Im \left[\psi^*(x) \frac{d}{dx} \psi(x) \right] =$$

$$= \frac{\hbar}{m} \Im \left[(e^{-ikx} + R^* e^{ikx})(ike^{ikx} - ikRe^{-ikx}) \right] =$$

$$= \frac{\hbar k}{m} \Re \left[(e^{-ikx} + R^* e^{ikx})(e^{ikx} - Re^{-ikx}) \right] =$$

$$= \frac{\hbar k}{m} \Re \left(1 - |R|^2 + R^* e^{2ikx} + Re^{-2ikx} \right) =$$

$$= \frac{\hbar k}{m} \left(1 - |R|^2 \right) = v \cdot 1 - v \cdot |R|^2.$$

The probability density current is made up of two terms constant in time: the first one represents a current of $1 \cdot v$ particles per second going through each point from left to right with speed v and the second one a current of $|R|^2 \cdot v$ particles per second propagating with opposite speed.

Region II: $x > 0$
In this region, we write the Schrödinger equation as:

$$-\frac{\hbar^2}{2m} \frac{d^2 \psi(x)}{dx^2} + V_0 \psi(x) = E \psi(x),$$

that is,

$$\frac{d^2 \psi(x)}{dx^2} + k'^2 \psi(x) = 0,$$

where

$$k' = \sqrt{\frac{2m(E - V_0)}{\hbar^2}} = \frac{p'}{\hbar} = \frac{mv'}{\hbar}.$$

We distinguish two cases, depending on whether the energy E m is greater than or less than V_0.

Case $E > V_0$
In this case, k' is real. Once again, the general solution is of the type

$$\psi(x) = Te^{ik'x} + Se^{-ik'x}, \tag{2.2}$$

where T and S are constants. Since the particle comes from the left, it must be $S = 0$. We are not calculating, therefore, the general solution, but rather a particular solution corresponding to particles sent from left to right. We therefore consider only the term

$$\psi(x) = Te^{ik'x}, \tag{2.3}$$

representing a probability density current of $v' \cdot |T|^2$ particles per second going through each point from left to right with speed v. To determine the coefficients R

and T, we use the continuity conditions for the wave function and its first derivative at the border between the two regions, the point $x = 0$. Thus, we obtain the linear system

$$\begin{cases} 1 + R = T \\ \imath k(1 - R) = \imath k'T. \end{cases}$$

The solution is

$$R = \frac{k - k'}{k + k'}, \tag{2.4}$$

$$T = \frac{2k}{k + k'}. \tag{2.5}$$

Substituting R and T in the expression for the current, we obtain

$$v|R|^2 = v\left(\frac{k - k'}{k + k'}\right)^2$$

$$v'|T|^2 = v'\frac{4k^2}{(k + k')^2}.$$

It is easy to verify that

$$v(1 - |R|^2) = v'|T|^2,$$

namely, the probability density current is the same in the two regions. Note that:

- In the limit $E \gg 0$, that is, for $k \longrightarrow k'$, we find $|R|^2 \longrightarrow 0$ and $|T|^2 \longrightarrow 1$, confirming the intuition that, at high energy, the potential step can be neglected leading to the free particle motion.
- An important novelty emerges with respect to the classical behavior of the particles: the presence of a perturbation in the potential generates a finite probability of reflecting back the particle.[1]

Case $E < V_0$
We note that, being that $E - V_0 < 0$, it results that

$$k'^2 = \frac{2m(E - V_0)}{\hbar^2} = (\imath \chi)^2, \quad \text{where} \quad \chi = \sqrt{\frac{2m(V_0 - E)}{\hbar^2}} \quad \text{is real and positive.}$$

We can therefore rewrite the Schrödinger equation as:

$$\frac{d^2\psi(x)}{dx^2} - \chi^2\,\psi(x) = 0. \tag{2.6}$$

---

[1] We see a similar behavior in optics when a change in the refractive index generates a reflected wave. In fact, monochromatic electromagnetic waves, once the time dependence has been removed, are ruled by the same equation, where the term $2m(V - E)/\hbar^2$ is replaced by k^2/c^2, where c is the propagation speed, function of the refractive index.

Its general solution is still of the type (2.2), but now the exponents are real:

$$\psi(x) = Te^{-\chi x} + Se^{\chi x}. \tag{2.7}$$

The second term must be eliminated, because it diverges in the limit $x = +\infty$; so, we have

$$\psi(x) = Te^{-\chi x}. \tag{2.8}$$

As in the previous case, we impose the continuity conditions and solve the resulting linear system or, faster, we substitute $\imath k' \to -\chi$ in (2.4) and (2.5). So we get the coefficients R and T:

$$R = \frac{k - \imath\chi}{k + \imath\chi}$$

$$T = \frac{2k}{k + \imath\chi}.$$

In this case, we note that

- R has modulus 1 and, because in region II the wave function is real, the transmitted current is zero; consequently, the reflected current is equal and opposite to the incident one and all of the particles, as in classical mechanics, go back.
- Contrary to what happens in classical mechanics, for which there can be no particles in region II because the kinetic energy would be negative, we have a non-zero probability of finding particles in the $x > 0$ region.

In both cases, $E > V_0$ and $E < V_0$, the energy eigenvalue can take any positive value, so the energy spectrum is **continuous**. However, in the two cases, the degeneration of the eigenvalues is different:

$E > V_0$. The spectrum is doubly degenerate. In fact, even if we found only one solution to the case, this is the consequence of having placed the coefficient $S = 0$, in order to reproduce a physically reproducible situation. There is, however, a solution that is linearly independent from this, corresponding to the sending of particles in the opposite direction to the x axis.

$E < V_0$. In this case, we put $S = 0$, because otherwise we would have had a solution divergent at infinity. One of the two linearly independent solutions to the eigenvalue equation is not in the wave function space and, consequently, eigenvalues between 0 and V_0 are not degenerate.

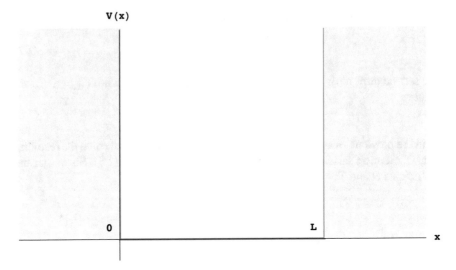

V(x)

0   L   **x**

Fig. 2.2 Particle confined on a segment (infinite potential well)

2.3 Particle Confined on a Segment (I)

A particle of mass m confined on a segment can be described by the potential (Fig. 2.2)

$$V(x) = \begin{cases} 0, & \text{if } 0 < x < L; \\ +\infty, & \text{elsewhere.} \end{cases}$$

also called the infinite potential well or one-dimensional box. Find the energy spectrum and the corresponding eigenfunctions.

Solution

The wave function is different from zero only for $0 < x < L$. Indeed, let's consider, for example, the $x > L$ region. Here, the potential is infinite; it is superior to any eigenvalue E of the energy we can fix. So, we can solve the Schrödinger equation by supposing that, in this region, the potential has a constant value $V_0 > E$ and then take the limit $V_0 \to \infty$ of the solution. We are in the same situation as in problem (2.2), when we consider the case $E < V_0$ for region II. We therefore have a single solution

$$\psi(x) = T e^{-\chi x}.$$

in the limit $V_0 \to \infty$, we get

$$\lim_{V_0 \to \infty} \chi = \lim_{V_0 \to \infty} \sqrt{\frac{2m(V_0 - E)}{\hbar^2}} = +\infty \qquad \Longrightarrow \qquad \lim_{V_0 \to \infty} \psi(x) = 0.$$

A similar reasoning can be repeated in the $x < 0$ region, and therefore the wave function is zero for $x > L$ and for $x < 0$.

Now we can solve the Schrödinger equation in the region $0 < x < L$ by requiring that the wave function go to zero at the ends of the segment. The general solution is

$$\psi(x) = A \sin(kx + \delta) \quad \text{where} \quad k = \sqrt{\frac{2mE}{\hbar^2}}.$$

The boundary condition in $x = 0$ leads to

$$\delta = 0,$$

and, applying the same condition in $x = L$, we get

$$kL = \sqrt{\frac{2mE}{\hbar^2}} L = n\pi \quad \text{where} \quad n = 1, 2, \ldots$$

From this relation, we obtain the energy eigenvalues

$$E_n = \frac{\hbar^2 \pi^2 n^2}{2mL^2} \quad \text{where} \quad n = 1, 2, \ldots \tag{2.9}$$

To determine the constant A, we normalize the eigenfunctions:

$$1 = \int_0^L dx\, |A|^2 \sin^2 \frac{n\pi x}{L} = \frac{|A|^2}{2} \int_0^L dx \left(1 - \cos \frac{2n\pi x}{L}\right) =$$
$$= \frac{|A|^2 L}{2} \quad \Longrightarrow \quad A = \sqrt{\frac{2}{L}},$$

up to an arbitrary phase factor. Therefore, the normalized eigenfunctions are

$$\psi_n(x) = \sqrt{\frac{2}{L}} \sin \frac{n\pi x}{L} \quad \text{with} \quad n = 1, 2, \ldots. \tag{2.10}$$

in the region $x \in [0, L]$ and zero elsewhere. These eigenfunctions are symmetrical with respect to $x = \frac{L}{2}$ for n odd and antisymmetrical for n even, due to the symmetry of the Hamiltonian. In conclusion, the energy spectrum is discrete and the eigenvalues are all non-degenerate.

2.4 Particle Confined on a Segment (II)

A particle of mass m moves in one dimension in the presence of an infinite well of width L:

$$V(x) = \begin{cases} 0 & \text{if } x \in [0, L], \\ +\infty & \text{elsewhere.} \end{cases}$$

Calculate the position and momentum expectation values and uncertainties when the particle is in an energy eigenstate. Comment on these results in light of the uncertainty principle.

Solution

Remember that, for the potential well (see problem 2.3), the energy eigenvalues are

$$E_n = \frac{n^2\pi^2\hbar^2}{2mL^2} \quad (n = 1, 2, \ldots) \tag{2.11}$$

and the corresponding eigenfunctions in the region $x \in [0, L]$ are given by

$$\psi_n(x) = \sqrt{\frac{2}{L}} \sin \frac{n\pi x}{L} \quad (n = 1, 2, \ldots) \tag{2.12}$$

and zero elsewhere. They have symmetry $(-1)^{n+1}$ (with respect to $x = \frac{L}{2}$), thus the probability distributions are always symmetrical, so

$$\langle x \rangle = \frac{L}{2}.$$

For a bound state, the momentum expectation value is always null (see problem 1.4). This general property can easily be demonstrated in the present case. Indeed, as the eigenfunctions are real (we can set the arbitrary phase factor equal to 1), we have

$$\langle p \rangle = \int_{-\infty}^{+\infty} dx\, \psi^*(x) \frac{\hbar}{\imath} \frac{d}{dx} \psi(x) = \frac{\hbar}{\imath} \frac{1}{2} \int_0^L dx \frac{d}{dx} \psi^2(x) = \frac{\hbar}{2\imath} \psi^2(x) \Big|_0^L = 0.$$

Let us calculate the momentum spread. As $\langle p \rangle = 0$,

$$(\Delta p)^2 = \langle p^2 \rangle = 2m\langle E \rangle = 2m\, E_n = \frac{n^2\pi^2\hbar^2}{L^2}.$$

The x^2 expectation value is given by

$$\begin{aligned} \langle x^2 \rangle &= \frac{2}{L} \int_0^L dx\, x^2 \sin^2 \frac{n\pi x}{L} = \frac{2L^2}{n^3\pi^3} \int_0^{n\pi} dy\, y^2 \sin^2 y = \\ &= \frac{2L^2}{n^3\pi^3} \left(\frac{n^3\pi^3}{6} - \frac{1}{4}n\pi \cos(2n\pi) - \frac{1}{8}(1 - 6n^2\pi^2)\sin(2n\pi) \right) = \\ &= L^2 \left(\frac{1}{3} - \frac{1}{2n^2\pi^2} \right), \end{aligned} \tag{2.13}$$

where we integrated repeatedly by parts.

The position uncertainty is given by

$$(\Delta x)^2 = \langle x^2 \rangle - \langle x \rangle^2 = L^2 \left(\frac{1}{3} - \frac{1}{2n^2\pi^2} \right) - \frac{a^2}{4} = L^2 \left(\frac{1}{12} - \frac{1}{2n^2\pi^2} \right).$$

The two uncertainty product is

$$\Delta x \Delta p = \hbar n \pi \sqrt{\frac{1}{12} - \frac{1}{2n^2\pi^2}}.$$

This product assumes its minimum value in the ground state (about $0.57\,\hbar$, slightly higher than $\hbar/2$) and grows as n increases.

2.5 Particle Confined on a Segment (III)

A particle of mass m, subject to the potential

$$V(x) = \begin{cases} 0 & \text{if } x \in [0, L], \\ +\infty & \text{elsewhere.} \end{cases}$$

is, at time $t = 0$, in the state corresponding to the wave function

$$\psi(x) = \frac{2}{\sqrt{L}} \cos \frac{\pi x}{2L} \sin \frac{3\pi x}{2L}.$$

(a) Write the wave function as a superposition of the Hamiltonian's eigenfunctions;
(b) calculate the energy expectation value;
(c) calculate the momentum expectation value;
(d) calculate the position expectation value.

Solution

Remember that, for the potential well (see problem 2.3), the energy eigenvalues are

$$E_n = \frac{n^2\pi^2\hbar^2}{2mL^2} \quad (n = 1, 2, \ldots) \tag{2.14}$$

and the corresponding eigenfunctions in the region $x \in [0, L]$ are given by

$$\psi_n(x) = \sqrt{\frac{2}{L}} \sin \frac{n\pi x}{L} \quad (n = 1, 2, \ldots) \tag{2.15}$$

and zero elsewhere.

(a) Taking into account the relationship

$$\cos\alpha\,\sin\beta = \frac{1}{2}[\sin(\alpha+\beta) - \sin(\alpha-\beta)],$$

we find

$$\psi(x) = \frac{1}{\sqrt{2}}\psi_1(x) + \frac{1}{\sqrt{2}}\psi_2(x).$$

(b) The energy expectation value is

$$\langle E\rangle_\psi = \langle\psi|\mathcal{H}|\psi\rangle = \frac{1}{2}E_1 + \frac{1}{2}E_2 = \frac{5\pi^2\hbar^2}{4mL^2}.$$

(c) In order to calculate the momentum expectation value, we recall that (see problem 2.4), for a particle in an energy eigenstate, it is zero. Therefore, we have

$$\langle p\rangle_\psi = \langle\psi|p|\psi\rangle = \frac{1}{2}\langle p\rangle_{\psi_1} + \frac{1}{2}\langle p\rangle_{\psi_2} + \frac{1}{2}\langle\psi_1|p|\psi_2\rangle + \frac{1}{2}\langle\psi_2|p|\psi_1\rangle =$$

$$= \frac{1}{2}\,2\,\Re(\langle\psi_1|p|\psi_2\rangle) = \Re\left(\frac{\hbar}{i}\frac{2}{L}\int_0^L dx\,\sin\frac{\pi x}{L}\cdot\frac{2\pi}{L}\cdot\cos\frac{2\pi x}{L}\right) = 0,$$

because it is the real part of a purely imaginary number.

(d) Finally, we calculate the position expectation value:

$$\langle x\rangle_\psi = \langle\psi|x|\psi\rangle = \frac{1}{2}\langle x\rangle_{\psi_1} + \frac{1}{2}\langle x\rangle_{\psi_2} + \frac{1}{2}\langle\psi_1|x|\psi_2\rangle + \frac{1}{2}\langle\psi_2|x|\psi_1\rangle =$$

$$= \frac{L}{2} + \frac{L}{2} + \langle\psi_1|x|\psi_2\rangle = a + \frac{2}{L}\int_0^L dx\,x\,\sin\frac{\pi x}{L}\,\sin\frac{2\pi x}{L} =$$

$$= a + \frac{4L}{\pi^2}\int_0^\pi d\alpha\,\alpha\,\sin^2\alpha\,\cos\alpha =$$

$$= a + \frac{4L}{3\pi^2}\left[\alpha\,\sin^3\alpha\Big|_0^\pi - \int_0^\pi d\alpha\,\sin^3\alpha\right] =$$

$$= a - \frac{4L}{3\pi^2}\left(z - \frac{z^3}{3}\right)\Big|_{-1}^{+1} = \left(1 - \frac{16}{9\pi^2}\right)L \simeq 0.82\,L.$$

2.6 Scattering by a Square-Well Potential

Consider the potential:

$$V(x) = \begin{cases} 0, & \text{if } |x| > a; \\ -V_0, & \text{if } |x| < a. \end{cases}$$

Consider a particle coming from the left with positive and study the behavior of the energy eigenfunctions and the features of its energy spectrum.

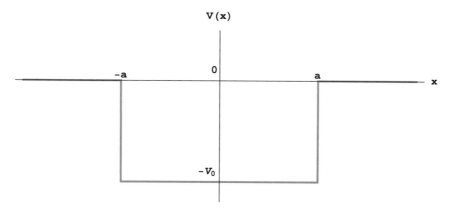

Fig. 2.3 Square-well potential

Solution

Having introduced the wave numbers k e k' for the regions in which the potential is constant

$$k = \sqrt{\frac{2mE}{\hbar^2}} \quad \text{and} \quad k' = \sqrt{\frac{2m(E + V_0)}{\hbar^2}},$$

we write the solution to the Schrödinger equation in the three constant potential regions as

$$\psi(x) = \begin{cases} e^{\imath kx} + Re^{-\imath kx} & \text{if } x < -a; \\ Ae^{\imath k'x} + Be^{-\imath k'x} & \text{if } |x| < a; \\ Te^{\imath kx}, & \text{if } x > a. \end{cases} \tag{2.16}$$

As requested, the solution represents the motion of a particle coming from the left that, interacting with the square-well, can be reflected or transmitted. The probability current in the three regions is

$$j(x) = \begin{cases} \frac{\hbar k}{m} - \frac{\hbar k}{m}|R|^2, & \text{if } x < -a; \\ \frac{\hbar k'}{m}|A|^2 - \frac{\hbar k'}{m}|B|^2, & \text{if } |x| < a; \\ \frac{\hbar k}{m}|T|^2, & \text{if } x > a. \end{cases}$$

The coefficients R, A, B and T, can be obtained by imposing the conditions of continuity of the wave function and its derivative at the border points between the regions of constant potential, a and $-a$. Thus, we obtain the following equations:

$$e^{-\imath ka} + Re^{\imath ka} = Ae^{-\imath k'a} + Be^{\imath k'a},$$
$$\imath k(e^{-\imath ka} - Re^{\imath ka}) = \imath k'(Ae^{-\imath k'a} - Be^{\imath k'a}),$$
$$Ae^{\imath k'a} + Be^{-\imath k'a} = Te^{\imath ka},$$
$$\imath k'(Ae^{\imath k'a} - Be^{-\imath k'a}) = \imath kTe^{\imath ka}.$$

Solving the system, we find the following expressions for R and T:

$$R = e^{-2\iota ka} \frac{(k'^2 - k^2) \sin 2k'a}{2kk' \cos 2k'a - \iota(k'^2 + k^2) \sin 2k'a}, \qquad (2.17)$$

$$T = e^{-2\iota ka} \frac{2kk'}{2kk' \cos 2k'a - \iota(k'^2 + k^2) \sin 2k'a}. \qquad (2.18)$$

Using these expressions, it is easy to verify that the probability current is the same in the three regions, i.e., it is conserved. We can conclude the study of this case with the following remarks:

- We note that

$$|T|^2 \propto (kk')^2 = \left(\frac{2m}{\hbar^2}\right)^2 (E + V_0) \qquad \text{while} \qquad |R|^2 \propto (k'^2 - k^2)^2 = \left(\frac{2m V_0}{\hbar^2}\right)^2.$$

Accordingly, in the limit of high energies ($E \gg V_0$), we have

$$kk' \sim \frac{2mE}{\hbar^2} \gg \frac{2m V_0}{\hbar^2} \implies |T|^2 \gg |R|^2,$$

and therefore reflection is negligible.
- Within the low energy limit, on the other hand, $T \longrightarrow 0$ and transmission is negligible.
- R is also proportional to $\sin 2k'a$, therefore, reflection is canceled (Resonance by Transmission) every time

$$k' = \frac{n\pi}{2a} \implies k'^2 = \frac{2m(E + V_0)}{\hbar^2} = \frac{n^2 \pi^2}{4a^2},$$

that is, for energy values

$$E_n = -V_0 + \frac{n^2 \pi^2 \hbar^2}{8ma^2}.$$

Regarding the properties of the energy spectrum, we conclude that it is continuous and twice degenerate.

2.7   Particle Confined in a Square-Well (I)

Consider the following potential well (vedi Fig. 2.3):

$$V(x) = \begin{cases} 0, & \text{if } |x| > a; \\ -V_0, & \text{if } |x| < a. \end{cases}$$

Determine the negative eigenvalues of the energy and the corresponding eigenfunctions.

Solution

In the two regions $|x| > a$, to the right and to left of the well, we are in the presence, as in the case $E < V_0$ for the potential step (see problem 2.2), of an equation of the kind in (2.6) which has the general solution

$$\psi(x) = c_1 e^{\chi x} + c_2 e^{-\chi x}, \qquad \text{where} \qquad \chi^2 = -\frac{2mE}{\hbar^2} > 0.$$

To avoid divergences at infinity of the wave function, in the left region, we will have to consider only the first of the two terms and, in the right region, only the second one. The wave function in the various regions has the form:

$$\psi(x) = \begin{cases} c_1 e^{\chi x}, & \text{if } x < -a; \\ A \cos k'x + B \sin k'x, & \text{if } |x| < a; \\ c_2 e^{-\chi x}, & \text{if } x > a, \end{cases}$$

where, in the central region, we have chosen as a general solution a combination of real functions (sines and cosines), instead of the complex exponentials of (2.16).

By imposing the continuity conditions of the wave function and its derivative, the following equation system is obtained:

$$c_1 e^{-\chi a} = A \cos k'a - B \sin k'a,$$
$$c_1 \chi e^{-\chi a} = Ak' \sin k'a + Bk' \cos k'a,$$
$$c_2 e^{-\chi a} = A \cos k'a + B \sin k'a,$$
$$-c_2 \chi e^{-\chi a} = -Ak' \sin k'a + Bk' \cos k'a.$$

We derive c_1 and c_2 from the first and third equations and replace their values in the second and fourth ones; we get two equations that allow us to calculate A and B:

$$\chi = k' \frac{A \sin k'a - B \cos k'a}{A \cos k'a + B \sin k'a}, \qquad (2.19)$$

$$\chi = k' \frac{A \sin k'a + B \cos k'a}{A \cos k'a - B \sin k'a}. \qquad (2.20)$$

If we equate the right members, we find that, in order for the two equations to be compatible, it must occur that

$$AB = 0.$$

We can therefore distinguish two types of solution,[2]

---

[2]The classification in odd and even eigenfunctions can be imposed a priori, since, for a potential invariant for parity, the eigenfunctions of the discrete spectrum have fixed parity.

- **Case B=0.** From the system, we see that $c_1 = c_2$ and the solution in the central region is $\psi(x) = A \cos k'x$. Therefore, considering the whole real axis, the solution is an **even** function of x.
- **Case A=0.** From the system, we see that $c_1 = -c_2$ and the solution in the central region is $\psi(x) = B \sin k'x$. Therefore, in this case, the solution is an **odd** function of x.

Even eigenfunctions. Imposing $B = 0$ in the Eqs. (2.19), we obtain

$$\chi = k' \tan k'a,$$

which, taking into account the relationship between χ and k'

$$\chi^2 = \frac{-2mE}{\hbar^2} = \frac{2mV_0}{\hbar^2} - k'^2, \tag{2.21}$$

can be rewritten as

$$\frac{\sqrt{\lambda^2 - z^2}}{z} = \tan z, \tag{2.22}$$

where

$$\lambda^2 = \frac{2mV_0a^2}{\hbar^2} \quad \text{and} \quad z = k'a$$

are both positive quantities. Energy E, in terms of the new variable z, is given by

$$E = \frac{\hbar^2 z^2}{2ma^2} - V_0. \tag{2.23}$$

As (2.22) is a transcendental equation, we can only qualitatively study the solutions using a graph in which we plot the first and second members as a function of the variable z. The z values of the intersection points between the two curves allow us to determine the eigenvalues of the energy through the relationship (2.21). In Fig. 2.4, we show the two members of Eq. (2.22) versus z. Function $\frac{\sqrt{\lambda^2 - z^2}}{z}$ has been drawn for various values of the parameter λ. This function intersects the z-axis at the point $z_{max} = \lambda = \sqrt{\frac{2mV_0a^2}{\hbar^2}}$. We note that

- as λ increases, the number of intersections between the two curves, and therefore the number of the energy eigenvalues, increases, but, although λ is small, there is always at least one;
- in the limit $\lambda \to \infty$, i.e., $V_0 \to \infty$, the intersections move towards

$$z_n = (n + \frac{1}{2})\pi = (2n + 1)\frac{\pi}{2} \quad \text{with} \quad n = 0, 1, \ldots$$

recovering the results obtained for the infinite potential well when n is odd (taking into account the replacement $2a \to L$).

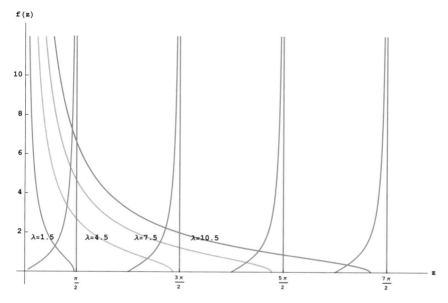

Fig. 2.4 Square well: graphical search for the even eigenfunctions' energy levels

Odd eigenfunctions. Imposing, instead, $A = 0$ in Eq. (2.19), we obtain the equation

$$\chi = -k' \cot k'a = k' \tan(k'a + \frac{\pi}{2}),$$

which, using the variables already defined, becomes

$$\frac{\sqrt{\lambda^2 - z^2}}{z} = \tan(z + \frac{\pi}{2}).$$

The solutions can be found, again, graphically, as we can see in Fig. 2.5. We note that

- for $\lambda < \pi/2$, we have no odd eigenfunction;
- as λ increases, the number of intersections, and therefore the number of the energy eigenvalues, increases;
- in the limit $\lambda \to \infty$, i.e., $V_0 \to \infty$, the intersections move towards

$$z_n = n\pi = (2n)\frac{\pi}{2} \quad \text{with} \quad n = 1, 2, \ldots$$

  recovering the results obtained for the infinite potential well when n is even.
- being antisymmetric, these solutions are zero in the origin, the same situation that one would have if the potential was given by

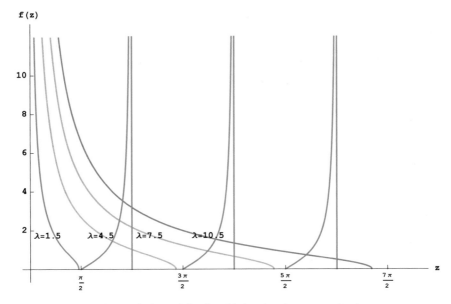

Fig. 2.5 Square well: graphical search for the odd eigenfunctions energy levels

$$V(x) = \begin{cases} \infty, & \text{if } x < 0, \\ -V_0, & \text{if } 0 < x < a. \end{cases}$$

So, this potential has energy eigenfunctions coinciding with the odd ones of the square well.

Looking at both figures, starting from $\lambda = 0$ and increasing λ, which is proportional to $V_0 a^2$, we first get an energy eigenvalue corresponding to an even eigenfunction, followed by the emergences of a second eigenvalue corresponding to an odd eigenfunction, and so on, alternating even solutions and odd solutions. This is a general behavior originating from the symmetry property of the potential. If

$$\lambda \in \left[(n-1)\frac{\pi}{2}, n\frac{\pi}{2} \right],$$

we have n energy levels and the eigenfunction corresponding to the n-th eigenvalue has parity $(-1)^{(n+1)}$.

We also note that, in each of the two cases ($B = 0$ and $A = 0$), the coefficients c_1 and c_2 are proportional to the other surviving coefficient (A or B, respectively). This residual coefficient can be fixed, up to the usual phase, by the normalization condition. Regarding the spectrum, we can say that within this range of energies, it is discrete, while the eigenfunctions go to zero exponentially at infinity; they are the quantum corresponding to the classical orbits limited to a region of space, i.e., bound states.

As we said, there is always a bound state. This result applies to any potential well $\bar{V}(x)$ of arbitrary form. In fact, it is always possible to bound $\bar{V}(x)$ above by a square potential $V(x)$; setting $\psi_0(x)$ as the energy eigenfunction of the ground state of $V(x)$, it results that

$$\left\langle \psi_0 \left| \frac{p^2}{2M} + \bar{V}(x) \right| \psi_0 \right\rangle < \left\langle \psi_0 \left| \frac{p^2}{2m} + V(x) \right| \psi_0 \right\rangle < 0,$$

and, in order to obtain an operator that has a negative expectation value, there must be at least one negative eigenvalue, that is a bound state.

2.8   Particle Confined in a Square-Well (II)

A particle of mass m moves in the one-dimensional potential (Fig. 2.6)

$$V(x) = \begin{cases} 0, & \text{if } |x| > a; \\ -V_0, & \text{if } |x| < a \end{cases}.$$

(a) At fixed width a, how deep should the well be in order to allow a first excited level of energy $E_1 = -\frac{1}{2}V_0$?
(b) If the particle is in the eigenstate of the Hamiltonian corresponding to the first excited level, what is the probability of finding it in the classically forbidden region?
(c) How many bound states of this system are there?

Solution

Remember that (see problem 2.7), once the following notations have been introduced,

$$\chi^2 = -\frac{2mE}{\hbar^2}, \quad k'^2 = \frac{2m}{\hbar^2}(V_0 + E),$$

the energy levels of the even eigenfunctions are obtained from

$$\chi = k' \tan k'a \tag{2.24}$$

and the energy levels for the odd ones from

$$\chi = -k' \cot k'a. \tag{2.25}$$

(a) Since $E_1 = -\frac{1}{2}V_0$, we have

$$\chi^2 = -\frac{2m}{\hbar^2}(-\frac{1}{2}V_0) = \frac{mV_0}{\hbar^2} = \frac{2m}{\hbar^2}\frac{1}{2}V_0 = k'^2.$$

Whereas the first excited level is odd, we have to find the smallest value of $k'a$ such that $\cot k'a = -\frac{\chi}{k'} = -1$, that is, $k'a = \frac{3}{4}\pi$. Therefore, the result is

$$V_0 = \frac{\hbar^2}{m}k'^2 = \frac{9\hbar^2}{16m}\frac{\pi^2}{a^2}.$$

(b) We seek the wave function of the first excited level having $k' = \chi = 3\pi/4a$. It must have the form

$$\psi_1(x) = \begin{cases} Ce^{\chi x}, & \text{if } x < -a; \\ B\sin k'x, & \text{if } |x| < a; \\ -Ce^{-\chi x}, & \text{if } x > a. \end{cases}$$

From the continuity condition in $x = a$

$$\begin{cases} B\sin k'a = -Ce^{-\chi a} \\ Bk'\cos k'a = C\chi e^{-\chi a} \end{cases}$$

(the two equations are equivalent because they are relative to an already fixed eigenvalue E_1) we get

$$\frac{B}{C} = \frac{\chi}{k'}\frac{e^{-\chi a}}{\cos k'a} = 1 \cdot \frac{e^{-\frac{3\pi}{4}}}{\cos\frac{3\pi}{4}} = -\sqrt{2}e^{-\frac{3\pi}{4}}.$$

Imposing the wave function normalization

$$\int_{-\infty}^{+\infty} |\psi_1(x)|^2 dx = 2\int_a^{+\infty}|C|^2 e^{-2\chi x}dx + \int_{-a}^a |B|^2\sin^2 k'x\, dx =$$

$$= \frac{|C|^2}{\chi}e^{-2\chi a} + |B|^2\int_{-a}^a \frac{1-\cos 2k'x}{2}dx =$$

$$= \frac{|C|^2}{\chi}e^{-2\chi a} + \frac{|B|^2}{2}\left(2a - \frac{1}{k'}\sin 2k'a\right) =$$

$$= 2ae^{-\frac{3\pi}{2}}\left(\frac{4}{3\pi}+1\right)|C|^2 = 1,$$

we deduce that

$$|C|^2 = \frac{3\pi}{4+3\pi}\frac{e^{\frac{3\pi}{2}}}{2a}.$$

Now we are able to calculate the probability P of finding the particle in the classically forbidden region as

$$P = 2\int_a^{+\infty}|C|^2 e^{-2\chi x}dx =$$

$$= \frac{3\pi}{4 + 3\pi} \frac{e^{\frac{3\pi}{2}}}{2a} \frac{e^{-2\chi a}}{\chi} =$$

$$= \frac{2}{4 + 3\pi}.$$

(c) Remember that (see problem 2.7), if

$$\lambda^2 = \frac{2m V_0 a^2}{\hbar^2} \in \left[(n-1)^2 \frac{\pi^2}{4}, n^2 \frac{\pi^2}{4} \right],$$

we have n energy levels and the eigenfunction corresponding to the n-th has parity $(-1)(n+1)$. The second excited level, corresponding to $n = 3$, is therefore obtained as the second solution of Eq. 2.24 for even eigenfunctions. In order for it to exist, it is necessary that $\lambda^2 = 2m V_0 a^2/\hbar^2 \geq \pi^2$. In the present case, this is verified:

$$\frac{2m}{\hbar^2} V_0 a^2 = \frac{2m}{\hbar^2} \frac{9\hbar^2}{16m} \pi^2 = \frac{9}{8}\pi^2 > \pi^2.$$

However, there is no second odd solution, as the relationship $\frac{2m}{\hbar^2} V_0 a^2 > \frac{9}{4}\pi^2$ is not verified.

In conclusion, there are only three bound states.

2.9 Potential Barrier

Consider the potential

$$V(x) = \begin{cases} 0, & \text{if } |x| > a; \\ V_0, & \text{if } |x| < a. \end{cases}$$

and suppose that particles of fixed energy $E < V_0$ strike it coming from $x = -\infty$. Determine the probability that a particle can cross the barrier (Fig. 2.6).

Solution

Notice that the energy eigenvalues, which must be greater than the potential minimum, are certainly positive; therefore $0 < E < V_0$. The solution to the Schrödinger equation for a particle coming from the left side is

$$\psi(x) = \begin{cases} e^{ikx} + Re^{-ikx}, & \text{for } x < -a; \\ Ae^{\chi x} + Be^{-\chi x}, & \text{for } |x| < a; \\ Te^{ikx}, & \text{for } x > a. \end{cases}$$

It is the same solution found for the square well (problem 2.6), but for the substitutions $V_0 \to -V_0$ and $ik' \to \chi$ ($\chi = \sqrt{\frac{2m(V_0-E)}{\hbar^2}}$ is real and positive). In fact, in the central region, the wave function has real exponents. We can, therefore, make the above

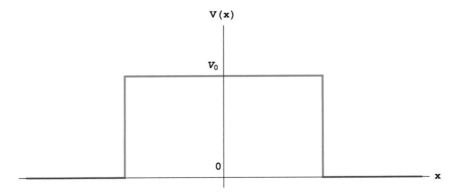

Fig. 2.6 Potential barrier

substitutions directly onto the results of problem 2.6. In particular, the transmission coefficient T is given by

$$T = e^{-2\imath ka} \frac{2k(-\imath\chi)}{2k(-\imath\chi)\cos 2(-\imath\chi)a - \imath((-\imath\chi)^2 + k^2)\sin 2(-\imath\chi)a},$$

and, taking into account the relationships

$$\cos \imath x = \cosh x \quad \text{and} \quad \sin \imath x = \imath \sinh x,$$

we obtain

$$T = e^{-2\imath ka} \frac{2k\chi}{2k\chi \cosh 2\chi a - \imath(k^2 - \chi^2)\sinh 2\chi a}.$$

The probability density current beyond the barrier is given by $\frac{\hbar k}{m}$, the velocity, times the square of the absolute value of T, which is

$$|T|^2 = \frac{(2k\chi)^2}{4k^2\chi^2 \cosh^2 2\chi a + (k^2 - \chi^2)^2 \sinh^2 2\chi a} =$$
$$= \frac{(2k\chi)^2}{4k^2\chi^2(1 + \sinh^2 2\chi a) + (k^2 - \chi^2)^2 \sinh^2 2\chi a} =$$
$$= \frac{(2k\chi)^2}{(k^2 + \chi^2)^2 \sinh^2 2\chi a + (2k\chi)^2}.$$

This expression, which also represents the ratio between the transmitted current and the incident one (which is equal to $\frac{\hbar k}{m}$), is certainly positive and tells us that, in Quantum Mechanics, contrary to the classical solution, there is always a probability of crossing (Tunnel effect) a potential barrier.

2.10 Particle Bound in a δ Potential

A particle of mass m moves in one dimension in the presence of the potential

$$V(x) = -\frac{\hbar^2}{m} \Omega \delta(x),$$

where $\delta(x)$ is the usual Dirac delta function. This system has a single bound state.

(a) Calculate the energy eigenvalue and the normalized eigenfunction of this state.
(b) Calculate the value x_0 such that the probability of finding the particle with $x < x_0$ is exactly equal to $1/2$.

Solution

Because we want bound states, we consider the eigenvalues $E < 0$ of the Schrödinger equation

$$\psi''(x) + 2\Omega\delta(x)\psi(x) - \alpha^2\psi(x) = 0, \quad \text{where} \quad \alpha^2 = -\frac{2mE}{\hbar^2} > 0.$$

(a) For $x \neq 0$, the solution satisfying the continuity condition in $x = 0$ is

$$\psi(x) = A\,e^{-\alpha|x|}.$$

Due to the presence of the δ potential, ψ' must be discontinuous in $x = 0$; otherwise its second derivative would assume a finite value and the δ singularity could not be compensated for in the Schrödinger equation. To find the discontinuity, we integrate between $-\epsilon$ and $+\epsilon$:

$$\psi'(x)\big|_{-\epsilon}^{+\epsilon} + 2\Omega \int_{-\epsilon}^{+\epsilon} dx\, \psi(x)\delta(x) + \alpha^2 \int_{-\epsilon}^{+\epsilon} dx\, \psi(x) = 0.$$

in the limit $\epsilon \to 0$, $\psi(x)$ being continuous in $x = 0$, we obtain

$$\psi'(0^+) - \psi'(0^-) = -2\Omega\psi(0). \tag{2.26}$$

This condition is satisfied by one value of α,

$$\alpha = \Omega. \tag{2.27}$$

There is therefore only one bound state having energy

$$E = -\frac{\hbar^2\Omega^2}{2m}.$$

The normalization condition sets the constant A up to a non-essential phase factor:

$$|A|^2 = \left[\int_{-\infty}^{0} e^{2\alpha x} \, dx + \int_{0}^{+\infty} e^{-2\alpha x} \, dx \right]^{-1} = \left[2(-\frac{1}{2\alpha}) e^{-2\alpha x} \Big|_{0}^{+\infty} \right]^{-1} = \alpha = \Omega.$$

The eigenfunction is given by

$$\psi(x) = \sqrt{\Omega} \, e^{-\Omega|x|}.$$

(b) The value x_0, which reduces the cumulative probability by half, is evidently 0, being that the wave function, and therefore also the probability distribution, is an even function.

2.11   Scattering by a δ Potential

A monochromatic beam of particles of mass m moves along the x axis in the presence of the potential

$$V(x) = -\frac{\hbar^2}{m} \Omega \, \delta(x),$$

where $\delta(x)$ is the Dirac delta function and Ω (opacity) is a positive quantity.

If the beam is incident from the left, a stationary wave function with energy E is given by

$$\psi(x) = \begin{cases} e^{ikx} + R e^{-ikx}, & \text{for } x \leq 0 \\ T e^{ikx}, & \text{for } x > 0. \end{cases}$$

with $k = \sqrt{2mE}/\hbar$.

Determine the probabilities of reflection and transmission beyond the barrier and study the limits of small and large opacity.

Solution

In order to determine the coefficients R and T at fixed energy E, we impose that, due to the presence of the δ potential, $\psi(x)$ is continuous and $\psi'(x)$ is discontinuous in $x = 0$ (see problem 2.10):

$$\psi(0^+) - \psi(0^-) = 0 \quad \Rightarrow \quad 1 + R = T$$
$$\psi'(0^+) - \psi'(0^-) = -2\Omega\psi(0) \quad \Rightarrow \quad ikT - ik(1 - R) = -2\Omega T.$$

From them, we get

$$R = \frac{i\Omega}{k - i\Omega} \quad \text{and} \quad T = \frac{k}{k - i\Omega}.$$

The probability flow transmitted through the barrier is given by

$$\frac{\hbar k}{m}|T|^2 = \frac{\hbar k}{m}\frac{1}{1+\frac{\Omega^2}{k^2}},$$

while the reflected flow is given by

$$-\frac{\hbar k}{m}|R|^2 = -\frac{\hbar k}{m}\frac{\frac{\Omega^2}{k^2}}{1+\frac{\Omega^2}{k^2}}.$$

Observe that the probability current is the same in the two regions

$$\frac{\hbar k}{m}|1|^2 - \frac{\hbar k}{m}|R|^2 = \frac{\hbar k}{m}|T|^2$$

and the currents, which are observable physical results, do not depend on the sign of Ω. In the limit $\Omega \to 0$, we have $|T|^2 \to 1$ and $|R|^2 \to 0$, and therefore all of the particles are transmitted beyond the δ barrier, while the reflection is zero. The opposite occurs in the limit $\Omega \to \infty$.

2.12 Particle Bound in a Double δ Potential

A particle of mass m moves in the one-dimensional potential

$$V(x) = -\frac{\hbar^2}{m}\Omega[\delta(x-a) + \delta(x+a)], \qquad \Omega > 0.$$

Prove that the Hamiltonian has, at most, two bound states and graphically solve the equation that determines them. Also, estimate the separation between levels for large values of a.

Solution

The Schrödinger equation becomes:

$$\frac{d^2\psi(x)}{dx^2} + 2\Omega[\delta(x-a) + \delta(x+a)]\psi(x) - \epsilon^2\psi(x) = 0 \quad \text{where} \quad \epsilon^2 = -\frac{2mE}{\hbar^2} > 0.$$

Remember that, due to the presence of the δ potential, ψ' must be discontinuous in $x = a$ and $x = -a$:

$$\psi'(a^+) - \psi'(a^-) = -2\Omega\psi(a),$$
$$\psi'(-a^+) - \psi'(-a^-) = -2\Omega\psi(-a).$$

As the potential is symmetrical, we can choose the solution with defined parity.

Let us first consider the **even** eigenfunctions.

In the three regions delimited by $x = a$ and $x = -a$, the Schrödinger equation has independent solutions given by

$$\psi_1(x) = e^{\epsilon x} \quad \text{and} \quad \psi_2(x) = e^{-\epsilon x} .$$

The solutions describing bond states have to go to zero at infinity and must be even; we can therefore write, disregarding an overall constant,

$$\psi_p(x) = \begin{cases} e^{\epsilon x}, & \text{for } x < -a; \\ A \cosh \epsilon x, & \text{for } |x| < a; \\ e^{-\epsilon x}, & \text{for } x > a. \end{cases}$$

Because of symmetry, it is sufficient to impose the continuity/discontinuity conditions at the point $x = a$ only:

$$\begin{cases} e^{-\epsilon a} = A \cosh \epsilon a \\ -A\epsilon \sinh \epsilon a - \epsilon e^{-\epsilon a} = -2\Omega e^{-\epsilon a}. \end{cases}$$

A solution for A is possible only if these two equations are compatible, i.e., only if

$$\tanh \epsilon a = \frac{2\Omega}{\epsilon} - 1 = \frac{2\Omega a}{\epsilon a} - 1. \tag{2.28}$$

The graphical solution is presented in Fig. 2.7, in which the two sides of Eq. (2.28) are plotted as a function of ϵa for various values of Ωa.

Let us now consider the **odd** eigenfunctions:

$$\psi_p(x) = \begin{cases} -e^{\epsilon x}, & \text{for } x < -a; \\ A \sinh \epsilon x, & \text{for } |x| < a; \\ e^{-\epsilon x}, & \text{for } x > a. \end{cases}$$

Imposing the continuity/discontinuity conditions in $x = a$,

$$\begin{cases} e^{-\epsilon a} = A \sinh \epsilon a \\ -A\epsilon \cosh \epsilon a - \epsilon e^{-\epsilon a} = -2\Omega e^{-\epsilon a}. \end{cases}$$

By imposing that the two equations for A are compatible, we get

$$\tanh \epsilon a = \frac{1}{\frac{2\Omega}{\epsilon} - 1} = \frac{1}{\frac{2\Omega a}{\epsilon a} - 1}. \tag{2.29}$$

The solution to this equation can be found graphically, as one can see in Fig. 2.8.

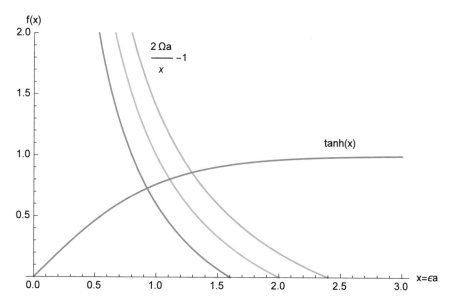

Fig. 2.7 Double δ potential: graphical solution of Eq. (2.28) for the even eigenfunctions. The right side has been drawn for $\Omega a = 0.8, 1.0, 1.2$

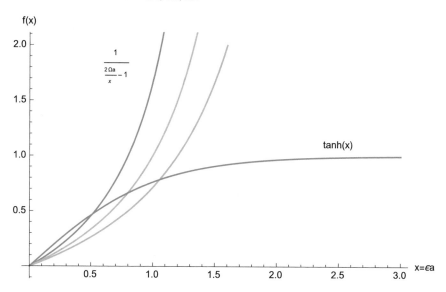

Fig. 2.8 Double δ potential: graphical solution of Eq. (2.29) for the odd eigenfunctions. The right side has been drawn for $\Omega a = 0.8, 1.0, 1.2$

A solution may exist, provided the slope in the origin of the function on the right in Eq. 2.29 is less than the slope of the function on the left, $\tanh \epsilon a$, which is 1:

$$\frac{d}{dx} \frac{x}{2\Omega a - x}\bigg|_{x=0} = \frac{1}{2\Omega a} < 1.$$

It is easy to see that, in this case, we have an excited state. In fact, the intersection, as $\tanh \epsilon a < 1$, is obtained for values of $\epsilon a < \Omega a$, while, for the even eigenfunctions, it was obtained for $\epsilon a > \Omega a$. So, for odd states, the corresponding energy $E = -\frac{\hbar^2}{2m}\epsilon^2$ is larger.

The separation between the two levels tends to zero in the limit of large distance between the two δ's. Indeed, the functions on the right side of Eqs. (2.28) and (2.29) are worth 1 in $\epsilon a = \Omega a$ and the function $\tanh \epsilon a$ also goes to 1 for large a.

2.13 Scattering by a Double δ Potential

Solve the Schrödinger equation for the potential

$$V(x) = \frac{\hbar^2}{m}\Omega(\delta(x - a) + \delta(x + a)), \quad \Omega > 0,$$

determining the Hamiltonian eigenvalues and eigenfunctions corresponding to a scattering problem. Discuss the energy dependence of the transmission coefficient.

Solution

As $V(x)$ is positive for every x, $\mathcal{H}$ eigenvalues are positive. Fixing $E = \hbar^2 k^2/2m > 0$, the E-eigenfunctions describing a particle sent in the positive direction of the x axis are

$$\psi_E(x) = \begin{cases} e^{ikx} + Re^{-ikx}, & \text{se } x < -a; \\ Ae^{ikx} + Be^{-ikx}, & \text{se } |x| < a; \\ Te^{ikx}, & \text{se } x > a, \end{cases}$$

where we set the probability current density equal to $|1|^2 \frac{\hbar k}{m}$. The continuity conditions for ψ and discontinuity conditions for its derivative ψ' (see problem 2.10),

$$\psi(\pm a^+) - \psi(\pm a^-) = 0$$
$$\psi'(\pm a^+) - \psi'(\pm a^-) = 2\Omega\psi(\pm a),$$

completely determine the coefficients R, A, B, T:

$$\begin{cases} e^{-ika} + Re^{ika} - Ae^{-ika} - Be^{ika} = 0 \\ Ae^{ika} + Be^{-ika} - Te^{ika} = 0 \\ (2\Omega + ik)e^{-ika} + (2\Omega - ik)Re^{ika} - ikAe^{-ika} + ikBe^{ika} = 0 \\ ikAe^{ika} - ikBe^{-ika} + (2\Omega - ik)Te^{ika} = 0. \end{cases}$$

It follows that every positive value of E is an eigenvalue of the Hamiltonian. Now defining

$$\alpha = \frac{ik - 2\Omega}{ik} = 1 + i\frac{2\Omega}{k} \quad , \quad \beta = e^{ika},$$

the II and IV equations become

$$\begin{cases} \beta A + \beta^* B = \beta T \\ \beta A - \beta^* B = \beta \alpha T. \end{cases}$$

So, we obtain

$$A = \frac{1}{2}(1 + \alpha)T \ , \quad B = \frac{\beta}{\beta^*}\frac{1}{2}(1 - \alpha)T.$$

Going back to the system, from the I and III equations, we obtain

$$\begin{cases} \beta^*(1 + \alpha^*) + \beta(1 - \alpha)R = \beta^*(1 + \alpha)T \\ \beta^*(1 - \alpha^*) + \beta(1 + \alpha)R = \frac{\beta^2}{\beta^*}(1 - \alpha)T, \end{cases}$$

from which, with simple steps, an expression for T is achieved:

$$T = \frac{1}{(1 + i\gamma)^2 + \gamma^2 e^{4ika}},$$

where $\gamma = \Omega/k$.

The transmission coefficient is $|T|^2$, which, with a few steps, becomes

$$|T|^2 = \frac{1}{(1 + \gamma^2)^2 + \gamma^4 + 2\gamma^2[(1 - \gamma^2)\cos 4ka + 2\gamma \sin 4ka]}.$$

We note that:

- $\lim_{k \to 0} |T|^2 = \lim_{\gamma \to \infty} |T|^2 = 0$, i.e., in the low energy limit the transmission coefficient goes to zero;
- $\lim_{k \to \infty} |T|^2 = \lim_{\gamma \to 0} |T|^2 = 1$, i.e., in the high energy limit the transmission is complete;
- $|T|^2$ shows oscillations corresponding to the behaviour of $(1 - \gamma^2)\cos 4ka + 2\gamma \sin 4ka$.

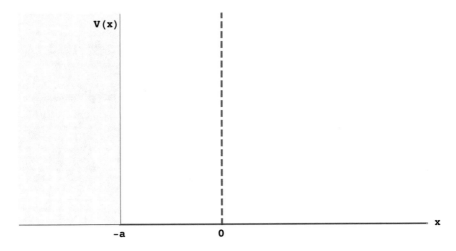

Fig. 2.9 δ potential in front of a wall

2.14 Collision Against a Wall in the Presence of a δ Potential

Consider a particle of mass m coming from $x = +\infty$ with energy $E > 0$ that bumps up against the potential (Fig. 2.9)

$$V(x) = \begin{cases} \infty, & \text{if } x \le -a; \\ \Omega\,\delta(x), & \text{if } x > -a. \end{cases}$$

(a) What happens in classical physics?
(b) Shape the wave function for $x < 0$ and for $x > 0$.
(c) Find the reflection probability.
(d) Find the reflected wave phase shift (with respect to the case of $\Omega = 0$) in the limit $x = +\infty$.
(e) Discuss the dependence from $\frac{\Omega}{k}$ $(k = \sqrt{\frac{2mE}{\hbar^2}})$ of the expressions found for the phase shift and for the wave function amplitude in section $-a < x < 0$.

Solution

(a) Classically, the particle, whatever its energy, would be reflected at position $x = 0$. This can be understood by thinking of the δ function as the limit of a rectangular function whose thickness tends to zero while its height tends to $+\infty$. The particle should have infinite energy in order to pass
(b) In the presence of this potential, the required E energy eigenfunction will take the form

$$\psi(x) = \begin{cases} A \sin(kx + \phi), & \text{if } -a \le x \le 0; \\ e^{-ikx} + R e^{ikx}, & \text{if } x > 0. \end{cases} \quad , \quad \text{with} \quad k = \sqrt{\frac{2mE}{\hbar^2}} \ ,$$

where we put the coefficient that represents the motion towards the barrier as being equal to one. The wave function must be zero in $x = -a$, so $\phi = ka$.

Let us set

$$\alpha = \frac{2m\Omega}{\hbar^2}, \tag{2.30}$$

which has dimension $[Length]^{-1}$, which is the same as k, because the Dirac δ also has dimension $[Length]^{-1}$ and Ω has dimension $[Energy][Length]$. The conditions for the wave function to be continuous and its derivative to be discontinuous in $x = 0$ lead to the system

$$\begin{aligned} \psi(0^+) = \psi(0^-) &\Rightarrow A \sin ka = 1 + R, \\ \psi'(0^+) - \psi'(0^-) = \alpha\psi(0) &\Rightarrow -kA \cos ka - ik(1 - R) = \alpha A \sin ka, \end{aligned}$$

which has the solution

$$\begin{aligned} A &= -\frac{2ik}{k \cos ka + \alpha \sin ka - ik \sin ka}, \\ R &= -\frac{k \cos ka + \alpha \sin ka + ik \sin ka}{k \cos ka + \alpha \sin ka - ik \sin ka}. \end{aligned} \tag{2.31}$$

The wave function is thus completely determined.

(c) As R is the ratio between two complex conjugate quantities, the reflection coefficient is given by

$$|R|^2 = 1.$$

As in the classical case, there is complete reflection, but the wave function is not zero between the δ barrier and the impenetrable wall.

(d) Calling ρ and θ, respectively, the module and the phase of the numerator of R, we have

$$R = e^{i\pi} \frac{\rho e^{i\theta}}{\rho e^{-i\theta}} = e^{i(2\theta + \pi)} \quad \text{where} \quad \theta = \arctan \frac{\tan ka}{1 + \frac{\alpha}{k} \tan ka}.$$

Now pose that $\Omega = 0$, i.e., $\alpha = 0$, in these formulas (absent the δ potential), obtaining

$$\theta \to \theta_0 = ka \quad \text{and} \quad R \to R_0 = e^{i(2ka + \pi)}.$$

The phase shift resulting from the barrier is therefore:

$$\Delta\varphi = 2\theta - 2\theta_0 = 2 \arctan \frac{\tan ka}{1 + \frac{\alpha}{k} \tan ka} - 2ka.$$

(e) Let us consider before the trend of $\Delta\varphi$ as a function of $\alpha/k = 2m\Omega/\hbar^2 k$. We note that:

- The phase shift $\Delta\varphi$ tends to 0 as α/k decreases.
- Apart from the trivial case of $\alpha = 0$, the phase shift $\Delta\varphi$ is zero if $\tan ka = 0$, that is, if $ka = n\pi$, with $n = 0, 1, 2, \ldots$. In these cases, the barrier becomes transparent.
- In the limit $\alpha/k \to +\infty$, $\Delta\varphi$ goes asymptotically to $-2ka$, corresponding to $R = -1$, that is, a situation in which the δ barrier becomes impenetrable. For this reason, the parameter Ω is often called *opacity*.

We now study the behavior for $x < 0$ of the wave function's A amplitude, or, better yet, of its square modulus:

$$|A|^2 = \frac{4k^2}{(k\cos ka + \alpha\sin ka)^2 + k^2\sin^2 ka} =$$
$$= \frac{4}{(\cos ka + \frac{\alpha}{k}\sin ka)^2 + \sin^2 ka}.$$

- At fixed k, $|A|^2$ takes the value 4 for $\alpha/k = 0$.
- Always considering k as fixed, $|A|^2$ is a descending function of α/k, thus confirming the role of Ω. It also has a maximum point at value $\alpha/k = -\cot ka$.
- The maximum is a resonance phenomenon, a phenomenon that can be better studied at fixed α. In Fig. 2.10, we note the presence of a peaks structure that is reduced when energy is increased.

2.15   Particle in the Potential $V(x) \propto -\cosh x^{-2}$

Consider a particle of mass m that moves in the one-dimensional potential

$$V(x) = -\frac{\hbar^2}{m}\frac{1}{\cosh^2 x}.$$

(a) Prove that
$$\psi(x) = (\tanh x + C)\exp(ikx)$$

is the solution to the Schrödinger equation for a particular value of the constant C. Determine such value and the energy corresponding to such a solution. From the asymptotical trend of $\psi(x)$, derive the reflection and transmission coefficients.

(b) In addition, show that
$$\phi(x) = \frac{1}{\cosh x}$$

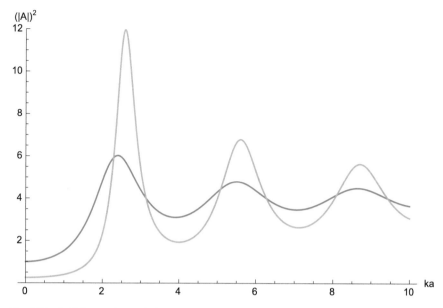

Fig. 2.10   Collision against an infinite barrier in the presence of a δ potential: the square amplitude $|A|^2$ of the transmitted wave as a function of ka for $\alpha = 1$ (blue curve) and $\alpha = 3$ (brown curve)

satisfies the Schrödinger equation. Show that it is a bound state and calculate its energy. Give an argument in favor of the fact that it is the ground state.

Solution

Having introduced

$$\varepsilon = \frac{2mE}{\hbar^2},$$

the Schrödinger equation becomes:

$$\frac{d^2}{dx^2}\psi(x) + \frac{2}{\cosh^2 x}\psi(x) + \varepsilon\psi(x) = 0.$$

(a) By imposing that $\psi(x)$ is one of its solutions, we find:

$$(\varepsilon - k^2)(\tanh x + C) + \frac{2}{\cosh^2 x}(ik + C) = 0.$$

This relation is verified for every x, provided that

$$\varepsilon = k^2 \quad \text{and} \quad C = -ik.$$

In the limit $x \to +\infty$,

$$\psi(x) \xrightarrow[x \to +\infty]{} (1 - ik)e^{ikx},$$

while, for $x \to -\infty$,

$$\psi(x) \xrightarrow[x \to +\infty]{} -(1 + ik)e^{ikx}.$$

There is therefore no reflected component ($\propto e^{-ikx}$):

$$R = 0 \quad \text{and} \quad T = 1.$$

(b) Regarding $\phi(x)$, by imposing that it is the solution to the Schrödinger equation, we obtain

$$-\frac{1}{\cosh x} + \frac{2}{\cosh x} + \frac{\varepsilon}{\cosh x} = 0,$$

and, therefore,

$$\varepsilon = -1.$$

Note that, for $|x| \to \infty \ \phi(x) \to 0$, and therefore $\phi(x)$ represents a bound state. Moreover, it is a function without nodes, and therefore it is the ground state.

2.16 Harmonic Oscillator: Position and Momentum

Calculate the matrix elements of position and momentum operators in the energy base of the harmonic oscillator. Evaluate the expectation values of both quantities in an energy eigenstate.

Solution

Using the expressions for operators x and p in terms of operators a and $a^\dagger$ (see A.14) and remembering that (see A.15)

$$a|n\rangle = \sqrt{n}\,|n-1\rangle, \, a^+|n\rangle = \sqrt{n+1}\,|n+1\rangle,$$

we have

$$x_{jk} = \langle j|x|k \rangle = \sqrt{\frac{\hbar}{2m\omega}}\,\langle j|(a + a^\dagger)|k\rangle = \sqrt{\frac{\hbar}{2m\omega}}\left[\sqrt{k}\,\delta_{k,j+1} + \sqrt{k+1}\,\delta_{k,j-1}\right],$$

$$p_{jk} = \langle j|p|k \rangle = \frac{1}{i}\sqrt{\frac{\hbar m\omega}{2}}\,\langle j|(a - a^\dagger)|k\rangle = -\iota\sqrt{\frac{\hbar m\omega}{2}}\left[\sqrt{k}\,\delta_{k,j+1} - \sqrt{k+1}\,\delta_{k,j-1}\right].$$

Regarding the expectation values, they are both null:

$$\langle x \rangle_k = \langle k|x|k\rangle = 0 \quad , \quad \langle p \rangle_k = \langle k|p\,|k\rangle = 0.$$

2.17   Harmonic Oscillator: Kinetic and Potential Energy

Calculate the matrix elements of operators x^2 and p^2 in the Harmonic oscillator energy basis. Show that, in an energy eigenstate, the expectation values of the kinetic energy and the potential energy are equal.

Solution

Using (A.14), we have

$$(x^2)_{jk} = \langle j|x^2|k \rangle = \frac{\hbar}{2m\omega} \langle j|a^2 + (a^\dagger)^2 + aa^\dagger + a^\dagger a|k \rangle, \tag{2.32}$$

$$(p^2)_{jk} = \langle j|p^2|k \rangle = -\frac{\hbar m\omega}{2} \langle j|a^2 + (a^\dagger)^2 - (aa^\dagger + a^\dagger a)|k \rangle. \tag{2.33}$$

From (A.15), we obtain

$$\langle j|a^2|k \rangle = \sqrt{k} \langle j|a|k-1 \rangle = \sqrt{k(k-1)}\,\delta_{k,j+2},$$

$$\langle j|(a^\dagger)^2|k \rangle = \sqrt{k+1} \langle j|a^\dagger|k+1 \rangle = \sqrt{(k+1)(k+2)}\,\delta_{k,j-2},$$

and, from $[a, a^\dagger] = 1$,

$$\langle j|(aa^\dagger + a^\dagger a)|k \rangle = \langle j|1 + 2a^\dagger a|k \rangle = \frac{2}{\hbar\omega} \langle j|\mathcal{H}|k \rangle = (2k+1)\delta_{j,k}.$$

Substituting these results in (2.32) and (2.33), we obtain the required matrix elements

$$(x^2)_{jk} = \frac{\hbar}{2m\omega} \left[\sqrt{k(k-1)}\,\delta_{k,j+2} + \sqrt{(k+1)(k+2)}\,\delta_{k,j-2} + (2k+1)\delta_{j,k} \right], \tag{2.34}$$

$$(p^2)_{jk} = -\frac{\hbar m\omega}{2} \left[\sqrt{k(k-1)}\,\delta_{k,j+2} + \sqrt{(k+1)(k+2)}\,\delta_{k,j-2} - (2k+1)\delta_{j,k} \right]. \tag{2.35}$$

Calling E_k the energy eigenvalue of the state $|k\rangle$, the required expectation values are given by

$$\langle x^2 \rangle_k = \langle k|x^2|k \rangle = \frac{\hbar}{2m\omega}(2k+1) = \frac{E_k}{m\omega^2}, \tag{2.36}$$

$$\langle p^2 \rangle_k = \langle k|p^2|k \rangle = \frac{\hbar m\omega}{2}(2k+1) = m E_k, \tag{2.37}$$

showing that the kinetic and potential energy, on average, are both equal to one half of the level energy. We have therefore recovered the result obtained by applying the Virial theorem (see problem 1.11).

A simpler method of calculation is to use the Feynman-Hellmann theorem (problem 1.12), which states that, in a steady state corresponding to the eigenvalue E, the following relation applies:

$$\left\langle \frac{\partial \mathcal{H}(\lambda)}{\partial \lambda} \right\rangle = \frac{\partial E(\lambda)}{\partial \lambda},$$

where λ is a parameter upon which the Hamiltonian and, therefore, also the eigenvalues and the eigenkets depend.

Applying this relation to a harmonic oscillator in the E_k state and using the frequency ω as the parameter, with respect to which to derive, we obtain:

$$\hbar \left(k + \frac{1}{2} \right) = \langle m\omega x^2 \rangle_k,$$

which implies

$$\langle x^2 \rangle_k = \frac{\hbar}{m\omega} \left(k + \frac{1}{2} \right) = \frac{E_k}{m\omega^2}.$$

Lastly, remembering that the eigenvalue E_k is the expectation value of the Hamiltonian in the kth state and that this is the sum of the expectation values of the kinetic energy and of the potential energy, the result for $\langle p^2 \rangle_k$ is also obtained.

2.18 Harmonic Oscillator: Expectation Value of x^4

Calculate the expectation value of operator x^4 in an energy eigenstate of the harmonic oscillator.

Solution

Using the completeness relation and the results of problem 2.17, we have

$$\langle x^4 \rangle_j = \langle j|x^4|j \rangle = \sum_{k=0}^{\infty} \langle j|x^2|k \rangle \langle k|x^2|j \rangle = \sum_{k=0}^{\infty} |\langle j|x^2|k \rangle|^2 = \sum_{k=0}^{\infty} \frac{\hbar^2}{4m^2\omega^2} \times$$
$$\times \left[\sqrt{k(k-1)}\, \delta_{k,j+2} + \sqrt{(k+1)(k+2)}\, \delta_{k,j-2} + (2k+1)\delta_{j,k} \right]^2.$$

Developing the square, the products of δ's with different indices do not contribute. So we get

$$\langle x^4 \rangle_j = \frac{\hbar^2}{4m^2\omega^2} \left[j(j-1) + (j+1)(j+2) + (2j+1)^2 \right] =$$
$$= \frac{3\hbar^2}{4m^2\omega^2} \left[2j^2 + 2j + 1 \right].$$

2.19 Harmonic Oscillator Ground State

A particle of mass m, subject to a harmonic oscillator of elastic constant k, is in its ground state. Calculate the probability of finding it outside the classically permitted region.

Solution

The classically permitted region is the segment between the classical turning points $\pm\bar{x}$, where

$$\bar{x} = \sqrt{\frac{2E}{k}}$$

is obtained solving the equation

$$V(\bar{x}) = \frac{1}{2}k\bar{x}^2 = E.$$

The ground state energy E is $\hbar\omega/2$ (with $\omega = \sqrt{k/m}$), so that $\bar{x} = \sqrt{\frac{\hbar}{m\omega}}$. The state is described by the wave function (see A.3)

$$\phi_0(x) = \left(\frac{m\omega}{\pi\hbar}\right)^{\frac{1}{4}} e^{-m\omega x^2/2\hbar}.$$

Taking into account the symmetry of the resulting probability distribution, the required probability is

$$P = 2\int_{\bar{x}}^{+\infty} dx \, |\phi_0(x)|^2 = \frac{2}{\sqrt{\pi}} \int_1^{+\infty} d\xi \, e^{-\xi^2} =$$

$$= \frac{2}{\sqrt{\pi}} \left[\frac{\sqrt{\pi}}{2} - \int_0^1 d\xi \, e^{-\xi^2}\right] = 1 - 2\,Erf(1) = 1 - 0.84 = 0.16, \quad (2.38)$$

where

$$Erf(y) = \frac{1}{\sqrt{\pi}} \int_0^y e^{-x^2} dx$$

is the Error function, which can be found in tables [1] or can be numerically evaluated.

2.20 Finding the State of a Harmonic Oscillator (I)

A harmonic oscillator of frequency ω is in a state superposition of the Hamiltonian eigenstates corresponding to the two lowest levels of energy:

$$|\psi\rangle = a|0\rangle + b|1\rangle.$$

(a) At what condition is the position expectation value different from zero?
(b) For which values of the coefficients a and b does this expectation value assume
    the maximum value and the minimum value?

Solution

Coefficients a and b must satisfy the normalization condition

$$|a|^2 + |b|^2 = 1.$$

We can fix an arbitrary phase and take a to be real:

$$|\psi\rangle = a|0\rangle + \sqrt{1 - a^2}\, e^{i\delta}\, |1\rangle.$$

(a) The position expectation value is, by (2.16),

$$\langle x\rangle = |a|^2\langle 0|X|0\rangle + |b|^2\langle 1|X|1\rangle + a^*b\langle 0|X|1\rangle + ab^*\langle 1|X|0\rangle =$$

$$= 2\Re(a^*b)\sqrt{\frac{\hbar}{2m\omega}} = 2\sqrt{\frac{\hbar}{2m\omega}}\, a\sqrt{1 - a^2}\, \cos\delta,$$

where we have taken into account that the position expectation value in an energy
eigenstate is zero.
For $\langle x\rangle$ to be different from zero, assuming that a and b are non-zero, it must be
that

$$\cos\delta \neq 0 \quad \Leftrightarrow \quad \delta \neq \left(n + \frac{1}{2}\right)\pi \quad \text{with } n \in \mathbb{N}.$$

(b) To determine the maximum and minimum of $\langle x\rangle$, we cancel the derivatives with
respect to the two parameters a and δ:

$$\frac{\partial\langle x\rangle}{\partial a} = 2\sqrt{\frac{\hbar}{2m\omega}}\frac{(1 - 2a^2)}{\sqrt{1 - a^2}}\cos\delta = 0,$$

$$\frac{\partial\langle x\rangle}{\partial\delta} = -2\sqrt{\frac{\hbar}{2m\omega}}\, a\sqrt{1 - a^2}\, \sin\delta = 0.$$

From the second condition, we get $\sin\delta = 0$, implying that $b = \pm\sqrt{1 - a^2}$.
The first condition cannot be satisfied by $\cos\delta = 0$, but only from $a = \pm\frac{1}{\sqrt{2}}$.
Ultimately, less than an arbitrary phase, we obtain

$$|\psi\rangle = \frac{1}{\sqrt{2}}|0\rangle \pm \frac{1}{\sqrt{2}}|1\rangle.$$

From the expression for $\langle x\rangle$, it is evident that it attains its maximum when the
sign is positive and its minimum when the sign is negative.

2.21 Finding the State of a Harmonic Oscillator (II)

It is known with certainty that the state of a harmonic oscillator of frequency ω contains no more excited states than the second level:

$$|\psi\rangle = a|0\rangle + b|1\rangle + c|2\rangle.$$

It is also known that the expectation value of the x position is zero and that the expectation value of the energy is $(3/4)\hbar\omega$.

What can be said of the values of a, b, c in the hypothesis that they are real? Is the state completely determined in these conditions?

Solution

Remembering that (A.14)

$$x|n\rangle = \sqrt{\frac{\hbar}{2m\omega}}(a + a^+)|n\rangle = \sqrt{\frac{\hbar}{2m\omega}}(\sqrt{n}|n-1\rangle + \sqrt{n+1}|n+1\rangle), \quad (2.39)$$

we obtain

$$\langle x\rangle = \sqrt{\frac{\hbar}{2m\omega}}(2ab + 2\sqrt{2}bc) = 0, \quad (2.40)$$

which has two solutions:

(a) $b \neq 0$ and $a = -\sqrt{2}c$,
(b) $b = 0$.

We have two other equations available:

$$a^2 + b^2 + c^2 = 1 \quad \text{(normalization)},$$
$$a^2 + 3b^2 + 5c^2 = \frac{3}{2} \quad \text{(energy expectation value)}.$$

In the first case, we obtain:

$$c = \pm\frac{\sqrt{3}}{2}, \quad b = \pm i\frac{\sqrt{5}}{2}, \quad a = \mp\sqrt{\frac{3}{2}},$$

which is not compatible with the hypothesis that coefficients are real.
  In the second case, we obtain:

$$b = 0, \quad c = \pm\frac{1}{2\sqrt{2}}, \quad a = \pm\sqrt{\frac{7}{8}}.$$

There are, finally, two possible determinations of the state, depending on whether a and c have a concordant or discordant sign.

Warning: The hypothesis of the reality of the coefficients, although useful for making the problem solvable, makes little physical sense, since the phase of each of them is not measurable. As a matter of fact, the results of this problem may depend on the definition used for the operators a and $a^\dagger$.

2.22 General Properties of Periodic Potentials

A periodic potential of step a is a potential $V(x)$ that enjoys the property

$$V(x + na) = V(x) \quad \text{for} \quad n = 0, \pm1, \pm2, \ldots. \tag{2.41}$$

In this case, the Schrödinger equation is invariant for transformations

$$x \rightarrow x + na,$$

i.e., for translations of integer multiples of a.

Given $u_1(x)$ and $u_2(x)$, two linearly independent solutions to the Schrödinger equation, due to the invariance property,

$$u_1(x + a) \quad \text{e} \quad u_2(x + a)$$

are also solutions. It must therefore result that

$$u_1(x + a) = c_{1,1}\, u_1(x) + c_{1,2}\, u_2(x) \tag{2.42}$$
$$u_2(x + a) = c_{2,1}\, u_1(x) + c_{2,2}\, u_2(x). \tag{2.43}$$

Prove:

(a) the Floquet theorem: among all of the solutions to the Schrödinger equation, there are two, ψ_1 and ψ_2, that satisfy the property

$$\psi(x + a) = \lambda \psi(x), \tag{2.44}$$

where λ is a constant;

(b) the Bloch theorem: such solutions can be written in the form

$$\psi(x) = e^{\imath k x} u_k(x),$$

where $u_k(x)$ is a periodic function of x of step a:

$$u_k(x + a) = u_k(x).$$

Solution

(a) Solutions of kind (2.44) must also satisfy

$$\psi(x + na) = \lambda^n \psi(x) \quad \text{for} \quad n = 0, \pm 1, \pm 2, \ldots \qquad (2.45)$$

If ψ is a solution, it is possible to write it in the form

$$\psi(x) = A u_1(x) + B u_2(x).$$

From (2.42), we have

$$\psi(x + a) = A u_1(x + a) + B u_2(x + a) =$$
$$= (Ac_{1,1} + Bc_{2,1}) u_1(x) + (Ac_{1,2} + Bc_{2,2}) u_2(x).$$

ψ verifies property (2.44) if

$$Ac_{1,1} + Bc_{2,1} = \lambda A,$$
$$Ac_{1,2} + Bc_{2,2} = \lambda B.$$

This is a homogeneous linear system of two equations in the variables A and B and has non-trivial solutions if and only if

$$\begin{vmatrix} c_{1,1} - \lambda & c_{2,1} \\ c_{1,2} & c_{2,2} - \lambda \end{vmatrix} = 0.$$

This is a second degree equation in λ, whose roots λ_1 and λ_2 allow us to actually determine two solutions ψ_1 and ψ_2 having the required property.

(b) We demonstrate first that λ_1 and λ_2 are complex conjugates numbers of modulus 1.
We note that the Wronskian of ψ_1 and ψ_2

$$W(x) = \psi_1 \psi_2' - \psi_1' \psi_2$$

satisfies the relation

$$W(x + a) = \lambda_1 \lambda_2 W(x).$$

As the Wronskian of two eigenfunctions corresponding to the same eigenvalue is constant (e.g., Messiah [11]. Vol.I, Chapter III.8), we have

$$\lambda_1 \lambda_2 = 1.$$

The functions ψ_1 and ψ_2 will be acceptable if and only if $|\lambda_1| = |\lambda_2| = 1$. Indeed, if $|\lambda_k| > 1$ happens, the amplitude ψ_k would grow beyond every limit for $x \to$

$+\infty$, while, if it were $|\lambda_k| < 1$, the same would happen for $x \to -\infty$. Therefore, we can put

$$\lambda_1 = e^{ika} \quad \text{and} \quad \lambda_2 = e^{-ika},$$

where k is a real number. As λ_1 and λ_2 are periodical functions, we can restrict ourselves to considering the values of k such that

$$-\frac{\pi}{a} \le k \le \frac{\pi}{a},$$

sufficient to determine all of the possible eigenfunctions. For any limited function you will therefore have (2.45):

$$\psi(x + n\,a) = e^{inka}\psi(x) \quad \text{for} \quad n = 0, \pm 1, \pm 2, \ldots.$$

This expression shows that the properties of invariance of the potential for translations of a length equal to step a are reflected on the wave function in such a way that, moving on an integer number of steps, it is modified only by a simple phase factor. It follows that the physical observables are not influenced by the translation, that is, the measure of any of them does not allow us to decide whether we are in x or $x + na$.

From this property, we deduce that, if we write $\psi(x)$ in the form

$$\psi(x) = e^{ikx}u_k(x),$$

$u_k(x)$ has to be a periodical function of x with step a:

$$u_k(x + a) = u_k(x). \tag{2.46}$$

This result is known as the *Bloch theorem*. It presents the eigenfunctions as a plane wave, the solution for a free particle system, with a modified amplitude reflecting the periodicity property of the potential.

2.23   The Dirac Comb

The simplest periodic potential that can be considered is the so-called Dirac comb:

$$V(x) = \frac{\hbar^2}{m}\,\Omega \sum_{n=-\infty}^{+\infty} \delta(x + na),$$

an infinite lattice of Dirac delta functions placed in positions $x = na$ with n integer. Determine the Bloch energy eigenfunctions (2.46) and show that the spectrum is composed of continuous bands of eigenvalues.

Solution

In each interval $]na, (n + 1)a[$, the particle is free, so, having fixed the energy eigen-value E, the plane waves are independent solutions of the Schrödinger equation:

$$u_1(x) = e^{\iota qx} \quad \text{and} \quad u_2(x) = e^{-\iota qx}, \quad \text{where} \quad q^2 = \frac{2mE}{\hbar^2}.$$

Now we look for the limited Floquet solutions (see problem 2.22), i.e., such that, if

$$\psi(x) = Ae^{\iota qx} + Be^{-\iota qx} \quad \text{in the interval} \quad 0 < x < a,$$

then it results that

$$\psi(x) = e^{\iota ka}\psi(x - a) = e^{\iota ka}[Ae^{\iota q(x-a)} + Be^{-\iota q(x-a)}] \quad \text{in the interval} \quad a < x < 2a.$$

We impose the continuity of the solution and, because there is a Dirac δ on the common endpoint (see (2.26)), the discontinuity of its first derivative:

$$\psi(a^+) = \psi(a^-),$$
$$\psi'(a^+) = \psi'(a^-) + 2\,\Omega\,\psi(a).$$

From these conditions, we obtain the linear system

$$e^{\iota ka}(A + B) = Ae^{\iota qa} + Be^{-\iota qa},$$
$$\iota qe^{\iota ka}(A - B) = \iota q(Ae^{\iota qa} - Be^{-\iota qa}) + 2\,\Omega\,(Ae^{\iota qa} + Be^{-\iota qa}),$$

that is,

$$(e^{\iota ka} - e^{\iota qa})A + (e^{\iota ka} - e^{-\iota qa})B = 0,$$
$$\left[e^{\iota ka} - e^{\iota qa}(1 - 2\iota\frac{\Omega}{q})\right]A - \left[e^{\iota ka} - e^{-\iota qa}(1 + 2\iota\frac{\Omega}{q})\right]B = 0.$$

This is a homogeneous system, so, to get non-trivial solutions, we have to impose the condition

$$\begin{vmatrix} e^{\iota ka} - e^{\iota qa} & e^{\iota ka} - e^{-\iota qa} \\ e^{\iota ka} - e^{\iota qa}(1 - 2\iota\frac{\Omega}{q}) & e^{\iota ka} - e^{-\iota qa}(1 + 2\iota\frac{\Omega}{q}) \end{vmatrix} = 0,$$

from which, with short steps, we get

$$\cos ka = \cos qa + \frac{\Omega}{q}\sin qa. \tag{2.47}$$

The first member must take values between -1 and 1, so we must have

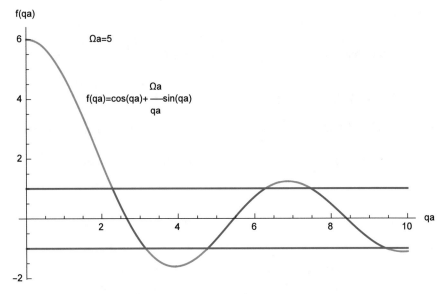

Fig. 2.11   Dirac comb: graphical solution to the inequality (2.48) for $\Omega a = 5$

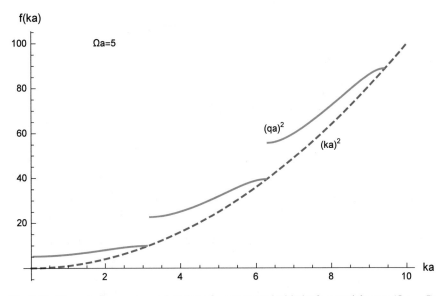

Fig. 2.12   Dirac comb: energy as a function of qa compared with the free particle case ($\Omega a = 5$)

$$\left| \cos qa + \frac{\Omega}{q} \sin qa \right| \le 1. \tag{2.48}$$

This inequality is verified for continuous energy intervals, bringing *bands* of eigenvalues into existence. Once we find the q values that satisfy inequality (2.48), in correspondence to each of them, we can determine, from Eq. (2.47), the k value characterizing the Bloch functions. Each energy level is

$$E = \frac{\hbar^2}{2m} q^2.$$

From Fig. 2.11, it can be noted that the forbidden bands, also called energy band gaps, are larger for small values of q and tend to cancel each other out in the limit $q \to \infty$. In Fig. 2.12, q^2, in practice the energy, is plotted as a function of k and compared with the parabola k^2 that one would have if the eigenfunctions were the plane waves, that is, if the motion were free and if q coincided with k. The two curves coincide at the points $k = \frac{n\pi}{a}$, which are the upper limit of each band.

It is interesting to study the influence of the parameter Ω, called opacity, on the energy spectrum. If Ω goes to zero, from Eq. (2.47), we see that $k \to q$, that is, the lattice becomes transparent and the forbidden bands more and more small until they vanish. If, conversely, $\Omega \to \infty$, the function on the right side of Eq. (2.47) takes on ever larger values and satisfies the condition (2.48) for ever more limited intervals of q. The allowed bands degenerate in the discrete spectrum that corresponds to the situation in which, in each segment of step a, there is a well with impenetrable walls.

2.24 The Kronig-Penney Model

The Kronig-Penney model consists of an infinite sequence of rectangular barriers with height of potential V_0, width b and separated by a distance $a - b$, so that a constitutes the lattice step (Fig. 2.13).

Determine the Hamiltonian's Bloch eigenfunctions and show that, for energies $E < V_0$, the spectrum is composed of continuous bands of eigenvalues.

Solution

To solve the model, we will use a slightly different method from the one used in the case of the Dirac comb (see problem 2.23).

During the demonstration of Floquet's Theorem (problem 2.22), we saw that

$$\lambda_1 = e^{\iota ka} \quad \text{and} \quad \lambda_2 = e^{-\iota ka} \tag{2.49}$$

are the eigenvalues of the matrix

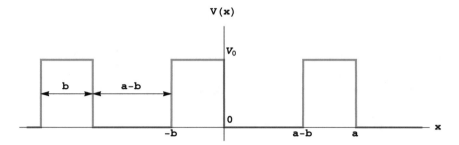

Fig. 2.13 Kronig-Penney model

$$\begin{pmatrix} c_{1,1} & c_{2,1} \\ c_{1,2} & c_{2,2} \end{pmatrix},$$

i.e., the matrix of the coefficients of the linear combination that expresses Floquet's wave functions in terms of any two solutions to the Schrödinger equation. Solving the eigenvalues equation, we find

$$\lambda_{1,2} = \frac{c_{1,1} + c_{2,2}}{2} \pm \sqrt{\left(\frac{c_{1,1} + c_{2,2}}{2} \right)^2 + c_{1,2}\, c_{2,1} - c_{1,1}\, c_{2,2}}.$$

This result, together with the expression (2.49) for λ_1 and λ_2, entails that

$$c_{1,1} + c_{2,2} = \lambda_1 + \lambda_2 = 2 \cos ka. \tag{2.50}$$

We now construct two linearly independent solutions $u_1(x)$ and $u_2(x)$, corresponding to an eigenvalue $E < V_0$. Having introduced the quantities

$$\alpha = \frac{\sqrt{2m(V_0 - E)}}{\hbar} \quad \text{and} \quad \beta = \frac{\sqrt{2mE}}{\hbar},$$

one solution is given by

$$u_1(x) = \mathrm{e}^{\alpha x} \quad \text{for } -b < x < 0,$$
$$u_1(x) = \cos \beta x + \frac{\alpha}{\beta} \sin \beta x \quad \text{for } 0 < x < a - b,$$

where we have set the coefficient of u_1 as equal to 1 and have imposed the continuity of the wave function and its derivative in $x = 0$.

In a completely similar way, we can determine an independent solution u_2 given by

$$u_2(x) = \mathrm{e}^{-\alpha x} \quad \text{for } -b < x < 0,$$
$$u_2(x) = \cos \beta x - \frac{\alpha}{\beta} \sin \beta x \quad \text{for } 0 < x < a - b.$$

In the region occupied by the next barrier, between $a - b$ and a, we must have

$$u_1(x) = c_{1,1}e^{\alpha(x-a)} + c_{1,2}e^{-\alpha(x-a)} \qquad \text{for } a - b < x < a,$$
$$u_2(x) = c_{2,1}e^{\alpha(x-a)} + c_{2,2}e^{-\alpha(x-a)} \qquad \text{for } a - b < x < a.$$

In $x = a - b$, the common endpoint for the two intervals, we can impose the continuity of $u_1(x)$, $u_2(x)$ and their derivatives leading to the system

$$\cos \beta(a - b) + \frac{\alpha}{\beta} \sin \beta(a - b) = c_{1,1}\,e^{-\alpha b} + c_{1,2}\,e^{\alpha b},$$

$$\cos \beta(a - b) - \frac{\alpha}{\beta} \sin \beta(a - b) = c_{2,1}\,e^{-\alpha b} + c_{2,2}\,e^{\alpha b},$$

$$-\beta \sin \beta(a - b) + \alpha \cos \beta(a - b) = \alpha(c_{1,1}\,e^{-\alpha b} - c_{1,2}\,e^{\alpha b}),$$

$$-\beta \sin \beta(a - b) - \alpha \cos \beta(a - b) = \alpha(c_{2,1}\,e^{-\alpha b} - c_{2,2}\,e^{\alpha b}),$$

from which it is possible to get the coefficients $c_{i,k}$. We obtain $c_{1,1}$ from the first and third equations, and $c_{2,2}$ from the second and fourth equations:

$$c_{1,1} = e^{\alpha b}\left[\cos \beta(a - b) + \frac{\alpha^2 - \beta^2}{2\alpha\beta} \sin \beta(a - b)\right],$$

$$c_{2,2} = e^{-\alpha b}\left[\cos \beta(a - b) - \frac{\alpha^2 - \beta^2}{2\alpha\beta} \sin \beta(a - b)\right].$$

Finally, using relationship (2.50), we get

$$\cos ka = \frac{c_{1,1} + c_{2,2}}{2} = \cosh \alpha b \, \cos \beta(a - b) + \frac{\alpha^2 - \beta^2}{2\alpha\beta} \sinh \alpha b \, \sin \beta(a - b).$$

$$(2.51)$$

As in the Dirac comb model, this relationship allows us to determine k only when the second member is between -1 and 1, thus giving rise to the band spectrum.

Chapter 3
Two and Three-Dimensional Systems

3.1 Plane Harmonic Oscillator

The Plane Harmonic oscillator has the Hamiltonian

$$H = \frac{1}{2m}(p_x^2 + p_y^2) + \frac{1}{2}m\omega^2(q_x^2 + q_y^2).$$

(a) Find the energy levels and their degeneracy;
(b) express the Hamiltonian in terms of the operators

$$\eta_+ = \frac{1}{\sqrt{2}}(a_x + ia_y) \qquad \eta_- = \frac{1}{\sqrt{2}}(a_x - ia_y),$$

where

$$a_x = \sqrt{\frac{m\omega}{2\hbar}}\,q_x + i\sqrt{\frac{1}{2m\omega\hbar}}\,p_x \qquad a_y = \sqrt{\frac{m\omega}{2\hbar}}\,q_y + i\sqrt{\frac{1}{2m\omega\hbar}}\,p_y,$$

and their Hermitian adjoint;
(c) write the angular momentum operator for this system; what can we say about the angular momentum at a fixed energy eigenvalue?

Solution

(a)
$$\mathcal{H} = \mathcal{H}_x + \mathcal{H}_y = \hbar\omega(a_x^\dagger a_x + a_y^\dagger a_y + 1).$$

$\mathcal{H}$ eigenvalues are given by

$$E = (n+1)\hbar\omega \quad \text{with } n = 0, 1, \ldots$$

© Springer Nature Switzerland AG 2019
L. Angelini, *Solved Problems in Quantum Mechanics*, UNITEXT for Physics,
https://doi.org/10.1007/978-3-030-18404-9_3

The corresponding eigenstates are $|n_x, n_y\rangle$ with $n_x + n_y = n$, $n_x \geq 0$, $n_y \geq 0$, and we can also write them as

$$|k, n - k\rangle \quad \text{con } k = 0, 1, \ldots, n.$$

E_n is therefore degenerate $n + 1$ times.

(b) In terms of η operators, we obtain

$$a_x = \frac{1}{\sqrt{2}}(\eta_+ + \eta_-) \quad a_y = \frac{1}{i\sqrt{2}}(\eta_+ - \eta_-)$$

$$\mathcal{H} = \hbar\omega(\eta_+^\dagger\eta_+ + \eta_-^\dagger\eta_- + 1).$$

(c) In this system, the angular momentum only has a component along the z axis. From

$$q_x = \sqrt{\frac{\hbar}{2m\omega}}(a_x + a_x^\dagger) \quad p_x = \frac{1}{i}\sqrt{\frac{\hbar m\omega}{2}}(a_x - a_x^\dagger),$$

we get

$$L = q_x p_y - q_y p_x = \frac{\hbar}{2i}\left[(a_x + a_x^\dagger)(a_y - a_y^\dagger) - (a_y + a_y^\dagger)(a_x - a_x^\dagger)\right] =$$
$$= \frac{\hbar}{i}\left[a_x^\dagger a_y - a_x a_y^\dagger\right].$$

In principle, it should be possible to find a set of simultaneous eigenkets for $\mathcal{H}$ and L, because one can easily show that these two operators commute. However, as required, we limit ourselves to studying the L matrix elements in the subspaces related to each energy eigenvalue, i.e., at n fixed. We obtain

$$\langle k', n - k'|L|k, n - k\rangle = \frac{\hbar}{i}\left(\sqrt{(k + 1)(n - k)}\,\delta_{k',k+1} - \sqrt{k(n - k + 1)}\,\delta_{k',k-1}\right),$$

where $k = 0, 1, \ldots, n$. We see immediately that the diagonal elements, which are the expectation values of L in the energy eigenstates we found before, are null. In these eigenstates, therefore, having called $\ell\hbar$, with ℓ relative integer or null, the eigenvalue of L, it occurs that $\ell = 0$, or there are combinations of ℓ and $-\ell$ that compensate. In each subspace relative to an E_n value, the matrix of L has this form:

$$
L^{(n)}_{k',k} =
\begin{pmatrix}
0 & \sqrt{n} & 0 & 0 & \cdots & 0 & 0 \\
-\sqrt{n} & 0 & \sqrt{2(n-1)} & 0 & \cdots & 0 & 0 \\
0 & -\sqrt{2(n-1)} & 0 & \sqrt{3(n-2)} & \cdots & 0 & 0 \\
0 & 0 & -\sqrt{3(n-2)} & 0 & \cdots & 0 & 0 \\
\vdots & \vdots & \vdots & \vdots & \vdots & \vdots & \vdots \\
0 & 0 & 0 & 0 & \cdots & 0 & \sqrt{n} \\
0 & 0 & 0 & 0 & \cdots & -\sqrt{n} & 0
\end{pmatrix}.
$$

$L^{(n)}$ is tridiagonal, antisymmetric with respect to the main diagonal, symmetric with respect to the antidiagonal. In general, it can be shown that the quantum number ℓ is subject to the condition

$$
\ell = -n, -n+2, \ldots, n-2, n.
$$

You can easily calculate the eigenvalues for the first values of n:

$$
\begin{aligned}
&\text{for } n = 0,\ \ell = 0, \\
&\text{for } n = 1,\ \ell = \pm 1, \\
&\text{for } n = 2,\ \ell = 0, \pm 2, \\
&\text{for } n = 3,\ \ell = \pm 1, \pm 3.
\end{aligned}
$$

3.2 Spherical Harmonic Oscillator

Consider a tridimensional harmonic oscillator with angular frequency ω. The Hamiltonian eigenkets depend upon three positive or null integers n_x, n_y, n_z:

$$
E(n_x, n_y, n_z) = \hbar\omega\left(n_x + n_y + n_z + \frac{3}{2}\right) = \hbar\omega\left(n + \frac{3}{2}\right),
$$

$$
|n\rangle = |n_x, n_y, n_z\rangle.
$$

(a) Express the angular momentum operators L_x, L_y, L_z as functions of the raising and lowering operators relevant to the various degrees of freedom, a_x, $a_x^\dagger$, a_y, $a_y^\dagger$, a_z, $a_z^\dagger$, and calculate the commutator of L_z with the Number operator for the z axis: $N_z = a_z^\dagger a_z$.
(b) Consider the three eigenstates relative to $n_x + n_y + n_z = 1$, determine their combinations that are eigenstates of L_z and calculate the L_z corresponding eigenvalues.

Solution

For each of the degrees of freedom $k = x, y, z$, we can define the operators

$$q_k = \sqrt{\frac{\hbar}{2m\omega}} \, (a_k + a_k^\dagger) \quad \text{and} \quad p_k = \frac{1}{i}\sqrt{\frac{m\omega\hbar}{2}} \, (a_k - a_k^\dagger),$$

$$a_k = \sqrt{\frac{m\omega}{2\hbar}} \, q_k + i\sqrt{\frac{1}{2m\omega\hbar}} \, p_k \quad \text{and} \quad a_k^\dagger = \sqrt{\frac{m\omega}{2\hbar}} \, q_k - i\sqrt{\frac{1}{2m\omega\hbar}} \, p_k$$

satisfying the relationships

$$[a_k, a_j] = [a_k^\dagger, a_j^\dagger] = 0 \quad , \quad [a_k, a_j^\dagger] = \delta_{k,j}.$$

(a) Let us calculate the angolar momentum components

$$L_x = q_y p_z - q_z p_y = \frac{\hbar}{2i}\left[(a_y + a_y^\dagger)(a_z - a_z^\dagger) - (a_z + a_z^\dagger)(a_y - a_y^\dagger)\right] =$$
$$= \frac{\hbar}{i}\left[a_y^\dagger a_z - a_y a_z^\dagger\right].$$

Similarly, it is found that

$$L_y = \frac{\hbar}{i}\left[a_z^\dagger a_x - a_z a_x^\dagger\right],$$

$$L_z = \frac{\hbar}{i}\left[a_x^\dagger a_y - a_x a_y^\dagger\right].$$

The commutator between L_z and N_z is

$$[L_z, N_z] = \frac{\hbar}{i}\left[a_x^\dagger a_y - a_x a_y^\dagger, N_z\right] = 0.$$

(b) The generic state corresponding to $n = n_x + n_y + n_z = 1$ is

$$|1\rangle = a|1, 0, 0\rangle + b|0, 1, 0\rangle + c|0, 0, 1\rangle.$$

Since we have seen that $[L_z, N_z] = 0$, we look for the eigenkets common to both operators. We immediately see that

- if $a = b = 0$ and $c = 1$, we have $N_z|1\rangle = N_z|0, 0, 1\rangle = 1 \cdot |0, 0, 1\rangle$.
  $|0, 0, 1\rangle$ is also an eigenket of L_z corresponding to the eigenket 0. Indeed,

$$L_z|0, 0, 1\rangle = \frac{\hbar}{i}\left[a_x^\dagger a_y - a_x a_y^\dagger\right]|0, 0, 1\rangle = 0.$$

- if $c = 0$ it results that $N_z|1\rangle = N_z[a|1, 0, 0\rangle + b|0, 1, 0\rangle] = 0$. That is, for any value of a and b, we have an eigenket of N_z corresponding to the eigenvalue 0. Let us search for which values of a and b this eigenket is also eigenket of

L_z, imposing that it satisfies the eigenvalues equation

$$L_z[a|1, 0, 0\rangle + b|0, 1, 0\rangle] = \frac{\hbar}{i}[-a|0, 1, 0\rangle + b|1, 0, 0\rangle] =$$
$$= m\hbar[a|1, 0, 0\rangle + b|0, 1, 0\rangle].$$

We immediately obtain

$$\begin{cases} -\frac{1}{i}a = m\,b \\ \frac{1}{i}b = m\,a \end{cases} \Rightarrow m^2 = \hbar^2 \quad \Rightarrow \quad m = \pm 1,$$

For $m = 1$, imposing the normalization condition $|a|^2 + |b|^2 = 1$, we find the eigenket

$$|n = 1, L_z = +\hbar\rangle = \frac{1}{\sqrt{2}}[|1, 0, 0\rangle + i|0, 1, 0\rangle],$$

while, for $m = -1$, we find the eigenket

$$|n = 1, L_z = -\hbar\rangle = \frac{1}{\sqrt{2}}[|1, 0, 0\rangle - i|0, 1, 0\rangle].$$

3.3 Reflection and Refraction in 3 Dimensions

Consider the three-dimensional step potential

$$V(x, y, z) = \begin{cases} 0, & \text{if } x < 0, \\ V_0, & \text{if } x > 0. \end{cases}$$

Derive the laws of reflection and refraction for a plane wave that impacts obliquely on the potential discontinuity and determine the conditions for total reflection.

Solution

The Schrödinger equation is separable in cartesian coordinates, because the Hamiltonian is given by:

$$\mathcal{H} = \begin{cases} \frac{p_x^2}{2m} + \frac{p_y^2}{2m} + \frac{p_z^2}{2m}, & \text{if } x < 0, \\ \frac{p_x^2}{2m} + V_0 + \frac{p_y^2}{2m} + \frac{p_z^2}{2m}, & \text{if } x > 0. \end{cases}$$

The system is invariant for rotations around the x axis; we can thus fix the incidence direction in the xy plane, putting $p_z = E_z = 0$. In this way, the problem involves only coordinates x and y. Because of separability, each eigenvalue has the form

$$E = E_x + E_y,$$

and the corresponding eigenfunctions are

$$\Psi(x, y) = \psi(x)\phi(y).$$

In the y coordinate, the motion is free, while, in the x coordinate, there is a step potential. Posing that

$$k_x = \sqrt{\frac{2m E_x}{\hbar^2}} \quad , \quad k_y = \sqrt{\frac{2m E_z}{\hbar^2}} \quad ,$$

we have, setting the coefficient of the incoming wave equal to 1,

$$\phi(y) = e^{ik_y y},$$

$$\psi(x) = \begin{cases} e^{ik_x x} + R e^{-ik_x x}, & \text{if } x < 0 \\ T e^{iq_x x}, & \text{if } x > 0 \end{cases}$$

where

$$q_x = \sqrt{\frac{2m(E_x - V_0)}{\hbar^2}}.$$

Thus, 3 wave vectors are identified:

- The incident wave vector $\mathbf{k} = (k_x, k_y)$ given by the boundary conditions;
- The reflected wave vector $\mathbf{k}' = (k'_x, k'_y) = (-k_x, k_y)$;
- The transmitted wave vector $\mathbf{q} = (q_x, k_y)$.

We denote with α, α' and β, respectively, the angles between these vectors and the x axis. We immediately find that the incidence angle and the reflection angle are the same, $\alpha = \alpha'$.

By imposing continuity conditions on the border of the two regions with different potential ($x = 0$), we obtain

$$R = \frac{k_x - k'_x}{k_x + k'_x} \quad \text{and} \quad T = \frac{2k_x}{k_x + k'_x}.$$

The situation presents similarities and differences with respect to the case of electromagnetic waves, for which Snell's law is valid. In the present case, V_0, which is the variation of the potential, assumes the role of a change in the refractive index. However, while for photons the simple relation is worth ($n =$ refractive index)

$$k = \frac{\omega}{c} n = \frac{E}{\hbar c} n,$$

for particles the relation is

$$k = \sqrt{\frac{2m(E - V_0)}{\hbar^2}}.$$

In terms of wave vectors, angles are given by

$$\sin\alpha = \frac{k_y}{\sqrt{k_y^2 + k_x^2}} \qquad \sin\beta = \frac{k_y}{\sqrt{k_y^2 + q_x^2}},$$

$$\frac{\sin\alpha}{\sin\beta} = \frac{\sqrt{k_y^2 + q_x^2}}{\sqrt{k_y^2 + k_x^2}} = \sqrt{\frac{E_y + E_x - V_0}{E_y + E_x}}.$$

We can distinguish two cases:

$V_0 > 0$. If $E < V_0$, q_x is imaginary and in the region $x > 0$ the wave function is exponentially damped. We find $|R| = 1$ and the transmitted probability current is zero: we have total reflection. If, instead, $E > V_0$, q_x is real and there is a transmitted probability current, we get $\sin\beta > \sin\alpha$ and, then, $\beta > \alpha$. This case corresponds, therefore, to the passage of light from a more refractive medium to a less refractive one.

$V_0 < 0$. Whatever the value of $E > 0$ we have $\alpha > \beta$. It's what happens when the light passes from a less refractive medium to a more refractive one.

3.4 Properties of the Eigenstates of J^2 and J_z

$J_\pm = J_x \pm iJ_y$ operators play the role of raising/lowering the z component of the angular momentum:

$$J_+|j, m\rangle = c_+|j, m + 1\rangle, \tag{3.1}$$

$$J_-|j, m\rangle = c_-|j, m - 1\rangle. \tag{3.2}$$

(a) Estimate the coefficients c_+ and c_- imposing the normalization of the eigenkets of J^2 and J_z.

(b) Calculate the matrix elements of operators J_x, J_y and J_z in the J^2, J_z basis.

(c) Prove that, in an eigenstate of J^2 and J_z corresponding to the quantum numbers j and m, the maximum accuracy of the simultaneous measurement of J_x and J_y is obtained when $|m| = j$.

(d) Calculate the expectation value, still in a J^2 and J_z eigenket, of the angular momentum component along a $\hat{n}$ direction that forms angle θ with the z axis.

Solution

(a) In the dual space, relation (3.1) becomes

$$\langle j, m | J_- = c_+^* \langle j, m+1 |,$$

and, therefore,

$$\langle j, m | J_- J_+ | j, m \rangle = |c_+|^2 \langle j, m+1 | j, m+1 \rangle = |c_+|^2,$$

from which we get

$$|c_+|^2 = \langle j, m | J_- J_+ | j, m \rangle = \langle j, m | J^2 - J_z^2 - \hbar J_z | j, m \rangle =$$
$$= [\hbar^2 j(j+1) - \hbar^2 m^2 - \hbar^2 m] \langle j, m | j, m \rangle.$$

Setting the c_+ phase equal to 0,

$$c_+ = \hbar \sqrt{j(j+1) - m(m+1)}. \tag{3.3}$$

In a similar way, we obtain

$$c_- = \hbar \sqrt{j(j+1) - m(m-1)}. \tag{3.4}$$

The relations (3.1) and (3.2) can, therefore, be written in a compact form:

$$J_\pm | j, m \rangle = \hbar \sqrt{j(j+1) - m(m \pm 1)} \, | j, m \pm 1 \rangle. \tag{3.5}$$

(b) Let us now calculate the matrix elements of **J** components in the basis $| j, m \rangle$,

$$\langle j', m' | J_x | j, m \rangle = \langle j', m' | \frac{J_+ + J_-}{2} | j, m \rangle =$$
$$= \frac{\hbar}{2} \Big[\sqrt{j(j+1) - m(m+1)} \, \delta_{j,j'} \delta_{m,m'-1} +$$
$$+ \sqrt{j(j+1) - m(m-1)} \, \delta_{j,j'} \delta_{m,m'+1} \Big], \tag{3.6}$$

$$\langle j', m' | J_y | j, m \rangle = \langle j', m' | \frac{J_+ - J_-}{2\imath} | j, m \rangle =$$
$$= \frac{\hbar}{2\imath} \Big[\sqrt{j(j+1) - m(m+1)} \, \delta_{j,j'} \delta_{m,m'-1} -$$
$$- \sqrt{j(j+1) - m(m-1)} \delta_{j,j'} \delta_{m,m'+1} \Big], \tag{3.7}$$

$$\langle j', m' | J_z | j, m \rangle = \hbar m \, \delta_{j,j'} \delta_{m,m'}, \tag{3.8}$$

(c) Let us now look for the minimum in the $| jm \rangle$ state of the uncertainties of J_x and J_y, which are the same for symmetry reasons (as, moreover, we can verify):

$$\langle\Delta J_x\rangle^2 = \langle J_x^2\rangle - \langle J_x\rangle^2 = \langle\Delta J_y\rangle^2 = \langle J_y^2\rangle - \langle J_y\rangle^2.$$

Again for reasons of symmetry: $\langle J_x\rangle = \langle J_y\rangle$. These expectation values vanish; in fact, using the previous results, we obtain:

$$\langle J_x\rangle = \langle j, m|J_x|j, m\rangle =$$
$$= \frac{\hbar}{2}\sqrt{j(j+1) - m(m+1)}\,\langle j, m|j, m+1\rangle +$$
$$+\frac{\hbar}{2}\sqrt{j(j+1) - m(m-1)}\,\langle j, m|j, m-1\rangle = 0.$$

Therefore,

$$\langle\Delta J_x\rangle^2 = \langle J_x^2\rangle = \frac{1}{2}\langle J^2 - J_z^2\rangle = \frac{1}{2}[j(j+1)\hbar^2 - m^2\hbar^2] =$$
$$= \frac{\hbar^2}{2}[j(j+1) - m^2],$$

which attains its minimum for $|m| = j$, which is the maximal value $|m|$ can assume.

(d) The versor $\hat{n}$ in spherical coordinates has the form

$$\hat{n} = (\sin\theta\cos\phi, \sin\theta\sin\phi, \cos\theta).$$

The **J** component along $\hat{n}$ is

$$\mathbf{J}\cdot\hat{n} = J_x\,\sin\theta\cos\phi + J_y\,\sin\theta\sin\phi + J_z\,\cos\theta.$$

The required expectation value is therefore

$$\langle j, m|\mathbf{J}\cdot\hat{n}|j, m\rangle = \sin\theta\cos\phi\langle j, m|J_x|j, m\rangle + \sin\theta\sin\phi\langle j, m|J_y|j, m\rangle +$$
$$+\cos\theta\langle j, m|J_z|j, m\rangle = \hbar m\cos\theta,$$

since, as we have seen, the expectation values for J_x and J_y are null.

3.5 Measurements of Angular Momentum in a State with $\ell = 1$

Consider a system in a $\ell = 1$ angular momentum state. After measuring the angular momentum component along a $\hat{n}$ direction that forms angle θ with the z axis and determining the result $+\hbar$, the angular momentum component along the z axis is measured. What is the probability of finding the value $+\hbar$?

Solution

The angular momentum components are compatible with its square modulus; therefore, their measures do not modify the property to be in the state with $\ell = 1$.

We will therefore develop the calculations in the basis $|\ell, m\rangle$, selecting the subspace relative to $\ell = 1$. As m can assume the values $+1, 0, -1$, a column array with 3 components corresponds to each ket and a 3×3 matrix to each operator.

Let us calculate the matrix elements of L_x, L_y and L_z using (3.6), (3.7), (3.8) for $j = \ell = 1$,

$$(L_x)_{m',m} = \langle 1, m' | L_x | 1, m \rangle =$$
$$= \frac{\hbar}{2} \left[\sqrt{2 - m(m+1)}\, \delta_{m,m'-1} + \sqrt{2 - m(m-1)}\, \delta_{m,m'+1} \right],$$
$$(L_y)_{m',m} = \langle 1, m' | L_y | 1, m \rangle =$$
$$= \frac{\hbar}{2\imath} \left[\sqrt{2 - m(m+1)}\, \delta_{m,m'-1} - \sqrt{2 - m(m-1)}\, \delta_{m,m'+1} \right],$$
$$(L_z)_{m',m} = \langle 1, m' | L_z | 1, m \rangle = m\hbar\, \delta_{m,m'}.$$

We can explicitly write these matrices if we associate the row or column index i with the value of m as follows:
$$m = +1,\ 0,\ -1,$$
$$i = \quad 1,\ 2,\ 3.$$

Thus, we get

$$L_x = \frac{\hbar}{\sqrt{2}} \begin{pmatrix} 0 & 1 & 0 \\ 1 & 0 & 1 \\ 0 & 1 & 0 \end{pmatrix}, \qquad L_y = \frac{\hbar}{\sqrt{2}} \begin{pmatrix} 0 & -\imath & 0 \\ \imath & 0 & -\imath \\ 0 & \imath & 0 \end{pmatrix}, \qquad L_z = \hbar \begin{pmatrix} 1 & 0 & 0 \\ 0 & 0 & 0 \\ 0 & 0 & -1 \end{pmatrix}.$$

It is easy to verify that L_x and L_y have the same eigenvalues as L_z, as expected, since all of the space directions are equivalent.

The L_z eigenvectors are

$$\psi_{+1} = \begin{pmatrix} 1 \\ 0 \\ 0 \end{pmatrix}, \qquad \psi_0 = \begin{pmatrix} 0 \\ 1 \\ 0 \end{pmatrix}, \qquad \psi_{-1} = \begin{pmatrix} 0 \\ 0 \\ 1 \end{pmatrix}.$$

The unit vector $\hat{n}$ is
$$\hat{n} = (\sin\theta \cos\phi, \sin\theta \sin\phi, \cos\theta)$$

and the **L** component along $\hat{n}$ is

$$\mathbf{L} \cdot \hat{n} = L_x \sin\theta \cos\phi + L_y \sin\theta \sin\phi + L_z \cos\theta,$$

i.e.,

$$\mathbf{L} \cdot \hat{n} = \hbar \begin{pmatrix} \cos\theta & \frac{1}{\sqrt{2}}\sin\theta e^{-i\phi} & 0 \\ \frac{1}{\sqrt{2}}\sin\theta e^{i\phi} & 0 & \frac{1}{\sqrt{2}}\sin\theta e^{-i\phi} \\ 0 & \frac{1}{\sqrt{2}}\sin\theta e^{i\phi} & -\cos\theta \end{pmatrix}.$$

The eigenvalues of $\mathbf{L} \cdot \hat{n}$ are, as for any component of $\mathbf{L}$, $+\hbar, 0, -\hbar$. We know that, after the measurement of $\mathbf{L} \cdot \hat{n}$, the system is in the eigenvector corresponding to the eigenvalue $+\hbar$, which we denote as

$$\psi_{+\hbar}^{\hat{n}} = \begin{pmatrix} a \\ b \\ c \end{pmatrix} = a\,\psi_{+1} + b\,\psi_0 + c\,\psi_{-1}.$$

Hence, the square modulus of a is the probability of finding $+\hbar$ in a subsequent measure of L_z. To determine $|a|^2$ we impose that $\psi_{+\hbar}^{\hat{n}}$ is the eigenvector for the eigenvalue $+\hbar$ di $\mathbf{L} \cdot \hat{n}$:

$$\mathbf{L} \cdot \hat{n}\, \psi_{+\hbar}^{\hat{n}} = \hbar\, \psi_{+\hbar}^{\hat{n}},$$

attaining the system

$$\begin{cases} a\cos\theta + \frac{b}{\sqrt{2}}\sin\theta e^{-i\phi} - a = 0 \\ \frac{a}{\sqrt{2}}\sin\theta e^{i\phi} + \frac{c}{\sqrt{2}}\sin\theta e^{-i\phi} - b = 0 \\ \frac{b}{\sqrt{2}}\sin\theta e^{i\phi} - c\cos\theta - c = 0 \end{cases}$$

Only two of these equations are independent, because this is an homogeneous system with zero coefficients determinant, but there is another equation as a result of the normalization condition:

$$|a|^2 + |b|^2 + |c|^2 = 1.$$

By developing short calculations, the required probability is found:

$$P(L_z = \hbar) = |a|^2 = \frac{(1 + \cos\theta)^2}{4}.$$

3.6 Angular Momentum of a Plane Wave

A particle has determinate momentum $\mathbf{p}$. Which result is obtained by measuring the angular momentum component along the $\mathbf{p}$ direction?

Solution

The particle wave function is a plane wave that propagates in the $\mathbf{p}$ direction:

$$\psi_{\mathbf{p}}(\mathbf{r}) = \left(\frac{1}{2\pi\hbar}\right)^{\frac{3}{2}} e^{\imath\mathbf{k}\cdot\mathbf{r}},$$

where $k = \frac{p}{\hbar}$.

By choosing the reference system so that $\mathbf{p}$ is directed along the z axis, one can write:

$$\psi_{\mathbf{p}}(\mathbf{r}) = \left(\frac{1}{2\pi\hbar}\right)^{\frac{3}{2}} e^{\imath kz} = \left(\frac{1}{2\pi\hbar}\right)^{\frac{3}{2}} e^{\imath kr\cos\theta}.$$

The $\mathbf{L}$ component along $\mathbf{p}$ is L_z, that is,

$$L_{\mathbf{p}} = L_z = -\imath\hbar\frac{\partial}{\partial\phi}.$$

As $\psi_{\mathbf{p}}$ does not depend on ϕ,

$$L_z\psi_{\mathbf{p}} = 0.$$

We can therefore state that the particle is in a $L \cdot \mathbf{p}$ eigenstate corresponding to eigenvalue 0.

3.7 Measurements of Angular Momentum (I)

The state of a particle of mass m is described by the wave function

$$\psi(\mathbf{r}) = \frac{1}{\sqrt{4\pi}}(e^{i\varphi}\sin\vartheta + \cos\vartheta)g(r),$$

where

$$\int |g(r)|^2 r^2 dr = 1$$

and φ, ϑ are the azimuth and polar angles, respectively.

(a) What are the possible results of a measurement of the L_z component of the particle angular momentum in this state?
(b) What is the probability of each of these possible results?
(c) What is the expectation value of L_z?

Solution

(c) Using the formula (A.39), the wave function can be rewritten in the form

$$\psi(\mathbf{r}) = \frac{1}{\sqrt{3}}(Y_{1,0} - \sqrt{2}Y_{1,1})g(r);$$

so, the possible values of L_z are $+\hbar$ and 0.

Assuming its radial part is normalized, the wave function is overall normalized. Indeed,

$$\int |\psi|^2 \, d\mathbf{r} = \int_0^\infty dr\, r^2 |g(r)|^2 \frac{1}{3} \int_{-1}^{+1} d\cos\vartheta \int_0^{2\pi} d\varphi \, |Y_{1,0} - \sqrt{2}Y_{1,1}|^2 =$$
$$= \frac{1}{3} \int_{-1}^{+1} d\cos\vartheta \int_0^{2\pi} d\varphi \, (|Y_{1,0}|^2 + 2|Y_{1,1}|^2) = 1.$$

So, the required probabilities are $P(L_z = \hbar) = 2/3$ e $P(L_z = 0) = 1/3$.
(c) $\langle L_z \rangle = 2/3 \cdot \hbar + 1/3 \cdot 0 = 2/3\,\hbar$.

3.8 Measurements of Angular Momentum (II)

A particle is in a state described by the wave function

$$\psi(x, y, z) = C(xy + yz + zx)e^{-\alpha r^2}.$$

(a) What is the probability that a measure of the square of angular momentum gives the result 0?
(b) What is the probability that a measure of the square of angular momentum gives the result $6\hbar^2$?
(c) If you find that the value of the orbital quantum number is 2, what are the probabilities related to the possible values of L_z?

Solution

By entering the spherical coordinates using the (A.21), we can write

$$\psi(r, \theta, \phi) =$$
$$= C\, r^2 e^{-\alpha r^2} (\sin^2\theta \, \sin\phi \, \cos\phi + \sin\theta \, \cos\theta \, \sin\phi + \sin\theta \, \cos\theta \, \cos\phi) =$$
$$= \frac{C}{2i} r^2 e^{-\alpha r^2} \left\{ \frac{1}{2} \sin^2\theta \, (e^{2i\phi} - e^{-2i\phi}) + \sin\theta \, \cos\theta \left[e^{i\phi}(1+i) - e^{-i\phi}(1-i) \right] \right\} =$$
$$= \frac{C}{2i} r^2 e^{-\alpha r^2} \sqrt{\frac{8\pi}{15}} \left[Y_{2,-2} - Y_{2,2} - (1+i)Y_{2,1} + (1-i)Y_{2,-1} \right],$$

where Spherical Harmonics were introduced using (A.40).

(a) The particle is in a state with $\ell = 2$, so $P(\ell = 0) = 0$.
(b) $P(L^2 = 6\hbar^2) = P(\ell = 2) = 1$.
(c) The probability of finding a certain value of L_z is given by the square modulus of the coefficient of the related Spherical Harmonic, after having integrated over

r and normalized the wave function. The result of the r-integral is the same for all of the components, so it is simpler to calculate the square modulus of the coefficients and then renormalize them to sum 1. The sum of the coefficients magnitudes is

$$1 + 1 + |1 + i|^2 + |1 - i|^2 = 1 + 1 + 2 + 2 = 6,$$

and the required probabilities are given by

$$P(L_z = -2\hbar) = \frac{1}{6},$$

$$P(L_z = +2\hbar) = \frac{1}{6},$$

$$P(L_z = -1\hbar) = \frac{1}{3},$$

$$P(L_z = +1\hbar) = \frac{1}{3},$$

$$P(L_z = 0\hbar) = 0.$$

3.9 Measurements of Angular Momentum (III)

A particle is in a state described by the wave function

$$\psi(\mathbf{r}) = A\, x\, e^{-\alpha r},$$

where α is a real constant and A is a normalization constant. If a measurement of L^2 is performed, what will be the wave function immediately after the measurement?

Solution

Remember that, from (A.39),

$$Y_{1,0} = \sqrt{\frac{3}{4\pi}} \cos\theta, \tag{3.9}$$

$$Y_{1,\pm 1} = \mp\sqrt{\frac{3}{8\pi}} \sin\theta\, e^{\pm i\phi}. \tag{3.10}$$

The cartesian coordinates,

$$x = r\,\sin\theta\,\cos\phi, \tag{3.11}$$

$$y = r\,\sin\theta\,\sin\phi, \tag{3.12}$$

$$z = r\,\cos\theta \tag{3.13}$$

can be written in terms of spherical harmonics. It follows that the wave function can, in turn, be rewritten in the form:

$$\psi(\mathbf{r}) = A\left[-\frac{r}{2}\left(\frac{8\pi}{3}\right)^{\frac{1}{2}}(Y_{1,+1} - Y_{1,-1})\right]e^{-\alpha r}.$$

It is therefore an eigenstate of L^2 corresponding to the eigenvalue

$$1(1+1)\hbar^2 = 2\hbar^2.$$

For this reason, the measurement of L^2 does not perturb the wave function.

3.10 Dipole Moment and Selection Rules

Consider the matrix elements of the position z component between the eigenstates $|\ell, m\rangle$ of L^2 and L_z:

$$\langle \ell, m | r \cos \theta | \ell', m' \rangle.$$

Using the Spherical Harmonics recurrence relation (A.35),

$$\cos \theta \, Y_{\ell,m}(\theta, \phi) = a_{\ell,m} \, Y_{\ell+1,m}(\theta, \phi) + a_{\ell-1,m} \, Y_{\ell-1,m}(\theta, \phi),$$

where

$$a_{\ell,m} = \sqrt{\frac{(\ell+1+m)(\ell+1-m)}{(2\ell+1)(2\ell+3)}}, \tag{3.14}$$

show that such matrix elements are zero unless the differences $\ell' - \ell$ and $m' - m$ assume particular values.

Solution

We calculate the required matrix elements in the spherical coordinates basis. Taking into account the suggested relationship, we have

$$\langle \ell, m | r \cos \theta | \ell', m' \rangle = r \int d\Omega \, Y_{\ell,m}^*(\theta, \phi) \cos \theta \, Y_{\ell',m'}(\theta, \phi) =$$

$$= r \int d\Omega \, Y_{\ell,m}^*(\theta, \phi) \left[a_{\ell',m'} \, Y_{\ell'+1,m'} + a_{\ell'-1,m'} \, Y_{\ell'-1,m'}\right] =$$

$$= r \, (a_{\ell-1,m} \, \delta_{\ell',\ell-1} + a_{\ell,m} \, \delta_{\ell',\ell+1}) \, \delta_{m,m'},$$

where we also used the orthogonal properties of the spherical harmonics (A.34). It follows that these matrix elements are null, unless

$$\Delta m = m' - m = 0 \quad \text{e} \quad \Delta \ell = \ell' - \ell = \pm 1. \tag{3.15}$$

These relationships are called selection rules, and they are relevant in calculating the probabilities of electric dipole transitions.

3.11 Quadrupole Moment

The quadrupole moment is the tensor

$$Q_{ik} = 3x_i x_k - r^2 \delta_{ik}.$$

Its expectation values indicate the deviations of the probability distribution compared to a spherically symmetrical one.

Determine its expectation values for one particle subject to a central potential in an eigenstate of $\mathcal{H}$, L^2 and L_z.

Solution

Because of the separability in spherical coordinates for a central potential, an eigenfunction of $\mathcal{H}$, L^2 and L_z assume the form (A.42), so that we have to calculate the following expressions:

$$\langle Q_{ik} \rangle = \int_0^\infty dr \, |\chi_\ell(r)|^2 \int_0^\pi d\theta \, \sin\theta \int_0^{2\pi} d\phi \, |Y_{\ell,m}(\theta, \phi)|^2 \, Q_{ik}.$$

The matrix elements Q_{ik} depend on the spherical coordinates through A.21. Consider non-diagonal elements: their integration function depends on ϕ via $\cos\phi$, $\sin\phi$ and $\sin\phi \cos\phi$. Therefore, all of these elements vanish because the integral over ϕ is null.

It remains to calculate the diagonal elements

$$Q_{11} = Q_{xx} = 3x^2 - r^2 = r^2(3\sin^2\theta \cos^2\phi - 1),$$
$$Q_{22} = Q_{yy} = 3y^2 - r^2 = r^2(3\sin^2\theta \sin^2\phi - 1),$$
$$Q_{33} = Q_{zz} = 3z^2 - r^2 = r^2(3\cos^2\theta - 1).$$

For Q_{xx} and Q_{yy}, the integral over ϕ, taking into account

$$\cos^2\phi = \frac{1 + \cos 2\phi}{2} \quad \text{and} \quad \sin^2\phi = \frac{1 - \cos 2\phi}{2},$$

gives us

$$\int_0^{2\pi} d\phi \, \cos^2\phi = \int_0^{2\pi} d\phi \, \sin^2\phi = \int_0^{2\pi} \frac{1}{2} \, d\phi.$$

Therefore, defining

$$\langle r^2 \rangle = \int_0^\infty dr \, r^2 \, |\chi_\ell(r)|^2,$$

we get

$$\langle Q_{xx} \rangle = \langle Q_{yy} \rangle = \langle r^2 \rangle \int_0^{2\pi} d\phi \int_0^\pi d\theta \, \sin\theta \left(\frac{3}{2} \sin^2\theta - 1 \right) |Y_\ell^m(\theta, \phi)|^2.$$

As

$$\frac{3}{2} \sin^2\theta - 1 = -\frac{1}{2}(3\cos^2\theta - 1),$$

we also get

$$\langle Q_{xx} \rangle = \langle Q_{yy} \rangle = -\frac{1}{2}\langle Q_{zz} \rangle.$$

We have only therefore to calculate $\langle Q_{zz} \rangle$. Taking into account the Spherical Harmonics recurrence relation (A.35), we obtain

$$\langle Q_{zz} \rangle = \langle r^2 \rangle \int_0^{2\pi} d\phi \int_0^\pi d\theta \, \sin\theta \, (3\cos^2\theta - 1) \, |Y_{\ell,m}(\theta, \phi)|^2 =$$

$$= \langle r^2 \rangle \left[3(a_{\ell,m}^2 + a_{\ell-1,m}^2) - 1 \right] = \langle r^2 \rangle \frac{2\ell(\ell+1) - 6m^2}{(2\ell-1)(2\ell+3)}.$$

Let's now look at some special cases.

 s wave: $\ell = 0$, $m = 0$ $\Rightarrow$ $\langle Q_{zz} \rangle = \langle Q_{xx} \rangle = \langle Q_{yy} \rangle = 0$, i.e., there is complete spherical symmetry.

 p wave, $\ell = 1$: here are three possibilities:

$m = 0$: $\Rightarrow$ $\langle Q_{zz} \rangle = \frac{4}{5}\langle r^2 \rangle > 0$, which indicates an elongated distribution in the z direction;

$m = \pm 1$: $\Rightarrow$ $\langle Q_{zz} \rangle = -\frac{2}{5}\langle r^2 \rangle < 0$, which indicates a flattened distribution in the z direction.

We notice that spherical symmetry is restored if we consider the sum of the $\langle Q_{zz} \rangle$ relative to the three states of m. This result is generally valid for every value of ℓ. Indeed,

$$\sum_{m=-\ell}^{+\ell} \langle Q_{zz} \rangle = \langle r^2 \rangle \frac{2}{(2\ell-1)(2\ell+3)} \sum_{m=-\ell}^{+\ell} [\ell(\ell+1) - 3m^2] = 0,$$

as

$$\sum_{m=-\ell}^{+\ell} m^2 = \frac{1}{3}\ell(\ell+1)(2\ell+1).$$

An atom that has electrons that complete the m states corresponding to the occupied ℓ states has spherical symmetry, and is therefore more stable with respect to electrical interactions (obviously, we are neglecting interaction between electrons).

3.12   Partial Wave Expansion of a Plane Wave

The Schrödinger equation in spherical coordinates for a free particle has the solutions

$$\psi_{\ell,m}(r, \theta, \phi) = a_{\ell,m}\, j_\ell(kr)Y_{\ell,m}(\theta, \phi). \tag{3.16}$$

These are called spherical waves and constitute an orthonormal basis of eigenvectors common to the compatible observables $\mathcal{H}$, L^2 and L_z. Instead, the Schrödinger equation in Cartesian coordinates has the solutions

$$\psi_{\mathbf{p}}(\mathbf{r}) = \left(\frac{1}{2\pi\hbar}\right)^{\frac{3}{2}} e^{i\mathbf{k}\cdot\mathbf{r}} \qquad \text{where } k = \frac{p}{\hbar},$$

which are called plane waves and make up an orthonormal basis of eigenvectors common to the compatible observables $\mathcal{H}$ and $\mathbf{p}$.

It must therefore be possible to express a plane wave as a superposition of spherical waves corresponding to the same energy $E = \frac{\hbar^2 k^2}{2m}$:

$$e^{i\mathbf{k}\cdot\mathbf{r}} = \sum_{\ell=0}^{\infty} \sum_{m=-\ell}^{+\ell} a_{\ell,m}\, j_\ell(kr)Y_{\ell,m}(\theta, \phi). \tag{3.17}$$

By choosing the z axis along the wave vector $\mathbf{k}$ direction, determine the coefficients of the development (3.17) by imposing that the two expressions have the same asymptotic behavior in r.

Solution

By choosing the z axis along the direction of the wave vector $\mathbf{k}$, the plane wave becomes $e^{ikr\cos\theta}$, which does not depend on ϕ. Also, the second member should not depend on ϕ and, since it is a superposition of linearly independent functions, this can only happen if only the terms with $m = 0$, which do not depend on ϕ, contribute to it. The spherical harmonics for $m = 0$ are given by

$$Y_{\ell,0}(\theta, \phi) = \sqrt{\frac{2\ell + 1}{4\pi}}\, P_\ell(\cos\theta),$$

and therefore (3.17) reduces to

$$e^{ikr\cos\theta} = \sum_{\ell=0}^{\infty} a_\ell \, j_\ell(kr) P_\ell(\cos\theta),$$

where all of the constants are incorporated into a_ℓ.

To determine the coefficients a_ℓ, we multiply both members for $P_\ell(\cos\theta)$ and we integrate over $\cos\theta$. Using the orthonormal relationship for Legendre Polynomials (A.30), we get

$$\int_{-1}^{+1} dz\, P_\ell(z) e^{ikrz} = \frac{2a_\ell}{2\ell+1} \, j_\ell(kr). \tag{3.18}$$

We now compare the r asymptotic behaviors of both sides. The series expansion of the first side of (3.18) can be obtained by iterating integration by parts:

$$\int_{-1}^{+1} dz\, P_\ell(z)\, e^{ikrz} = \frac{1}{ikr}\left[e^{ikrz} P_\ell(z)\right]_{-1}^{+1} - \frac{1}{ikr}\int_{-1}^{+1} dz\, P_\ell'(z)\, e^{ikrz} =$$

$$= \frac{1}{ikr}\left[e^{ikrz} P_\ell(z)\right]_{-1}^{+1} - \frac{1}{ikr}\left\{\frac{1}{ikr}\left[e^{ikrz} P_\ell'(z)\right]_{-1}^{+1} - \frac{1}{ikr}\int_{-1}^{+1} dz\, P_\ell''(z)\, e^{ikrz}\right\} =$$

$$= \frac{1}{ikr}\left[e^{ikrz} P_\ell(z)\right]_{-1}^{+1} + O(\frac{1}{r})^2 = \frac{1}{ikr}\left[e^{ikr} - e^{i\pi\ell} e^{-ikr}\right] + O(\frac{1}{r})^2 =$$

$$= \frac{2}{kr} e^{\frac{i\pi\ell}{2}} \sin\left(kr - \frac{\pi\ell}{2}\right) + O(\frac{1}{r})^2,$$

where we used the property (A.31) of Legendre polynomials.

To calculate the r asymptotic behavior of the second side of (3.18), we use (A.54), and then, from (3.18), we obtain

$$\frac{2}{kr} i^\ell \sin\left(kr - \frac{\pi\ell}{2}\right) = \frac{2a_\ell}{2\ell+1} \frac{1}{kr} \cos\left(kr - \frac{\ell+1}{2}\pi\right),$$

from which, simplifying, we reduce to the following expression for a_ℓ:

$$a_\ell = i^\ell \, (2\ell+1). \tag{3.19}$$

Thus, we obtain the spherical wave expansion of a plane wave:

$$e^{ikr\cos\theta} = \sum_{\ell=0}^{\infty} i^\ell \, (2\ell+1) \, j_\ell(kr) P_\ell(\cos\theta). \tag{3.20}$$

3.13 Particle Inside of a Sphere

Determine the eigenvalues of the discrete energy spectrum for a particle surrounded by a sphere of impenetrable potential, within which the potential is null.

Solution

We denote the radius of the sphere with R. Its impenetrability implies that the wave function vanishes on the surface, so that the outgoing probability current is zero. Inside the sphere, instead, the particle is free and, normalization apart, the radial wave function is given by

$$R_{k,\ell}(r) = j_\ell(kr),$$

where $k = \sqrt{2mE}/\hbar$ and $j_\ell(z)$ is the ℓth Bessel spherical function (see A.7). As already noted, the radial wave function must vanish on the boundary and therefore

$$j_\ell(kR) = 0.$$

The Bessel spherical functions depend on trigonometric functions and have infinite zeros that are numbered in ascending order. Having called $\bar{z}_{n_r,\ell}$ the n_r-simo zero of j_ℓ, in order for $j_\ell(kR)$ to be zero, it must result that $k = \frac{\bar{z}_{n_r,\ell}}{R}$. We conclude that the possible energy eigenvalues are given by

$$E_{n_r,\ell} = \frac{\hbar^2}{2m} \frac{\bar{z}_{n_r,\ell}^2}{R^2}. \tag{3.21}$$

n_r is called the radial quantum number, to distinguish it from the orbital quantum number ℓ. Zeros of the spherical Bessel functions are tabulated (see, for example, [1]).

3.14 Bound States of a Particle Inside of a Spherical Potential Well

Determine the energy eigenvalues of the bound states of a particle subject to the potential:

$$V(r) = \begin{cases} 0, & \text{per } r > a; \\ -V_0, & \text{per } r < a. \end{cases}$$

Solution

Taking into account centrifugal potential, the particle is in the presence of the effective potential:

$$V_{eff}(r) = V(r) + \frac{\hbar^2 \ell(\ell+1)}{2mr^2}, \tag{3.22}$$

which has been drawn in black in Fig. 3.1. The condition necessary for the bound states to exist is that the minimum of potential be lower than its asymptotic value, which is null. Since the minimum is always placed in $r = a$, this is equivalent to the condition

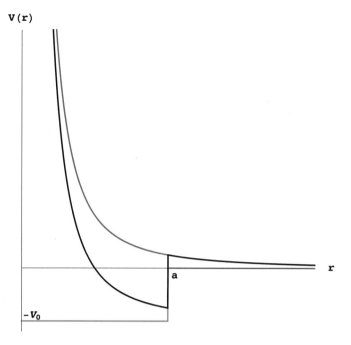

Fig. 3.1 Spherical well: the spherical potential well (blue), which added to the centrifugal potential (red), gives rise to the effective potential (black)

$$V_0 > \frac{\hbar^2 \ell(\ell + 1)}{2ma^2}. \tag{3.23}$$

We can already say that there are situations for which there have been no bound states and others for which they are present, but only for the lower states of angular momentum. We will see, however, that condition (3.23) is not a sufficient condition. At fixed values of ℓ and energy $E < 0$, we can distinguish two regions, depending on whether r is less or greater than a.

Region I: $r < a$
Inside the sphere, there is a potential constantly equal to $-V_0$. The radial equation is that of a free particle:

$$\frac{d^2}{dr^2} U_{k,\ell}(r) + \left[k^2 - \frac{\ell(\ell + 1)}{r^2}\right] U_{k,\ell}(r) = 0, \tag{3.24}$$

with wave number k

$$k^2 = \frac{2m(E + V_0)}{\hbar^2}.$$

Since the origin is included in this region, the only acceptable solution (satisfying the condition $\lim_{r \to 0} U_{k,\ell}(r) = 0$) is

$$U_{k,\ell}(r) = A\,r\,j_\ell(kr), \tag{3.25}$$

where A is a normalization constant.

Regione II: $r > a$
Outside of the sphere, the potential is also constant, but now its value is zero. The
radial equation is that of a free particle with eigenvalues that are, however, negative:

$$\frac{d^2}{dr^2}U_{k,\ell}(r) + \left[-\chi^2 - \frac{\ell(\ell+1)}{r^2}\right]U_{k,\ell}(r) = 0, \tag{3.26}$$

where χ is

$$\chi^2 = -\frac{2mE}{\hbar^2} > 0.$$

The solutions to this equation are still spherical Bessel functions, with the replacement $k \to \iota\chi$. The condition to be imposed on these solutions is not the regular
behavior in the origin, which is not part of this region, but rather that they do not
diverge for $r \to +\infty$. As explained in the Appendix, we see that the solutions with
correct asymptotic pattern are the spherical Hankel functions of first kind (A.62).
The solution in this region is, therefore, given by

$$U_{k,\ell}(r) = B\,r\,h_\ell^{(1)}(\iota\chi r). \tag{3.27}$$

Connecting the solutions
The solutions found in the two regions must be equal in $r = a$, the border point,
together with their derivatives, or, equivalently, together with their logarithmic derivatives. The continuity condition on the wave function determines the ratio B/A (the
normalization condition then allows us to determine the module of each of them).
The continuity condition on the logarithmic derivative, on the other hand, determines
the spectrum of the energy eigenvalues. In fact, it entails that

$$\frac{1}{h_\ell^{(1)}(\iota\chi a)}\left[\frac{d}{dr}h_\ell^{(1)}(\iota\chi r)\right]_{r=a} = \frac{1}{j_\ell(ka)}\left[\frac{d}{dr}j_\ell(kr)\right]_{r=a}. \tag{3.28}$$

Note that k and χ are not independent variables, because

$$\chi^2 = -\frac{2mE}{\hbar^2} = \frac{2mV_0}{\hbar^2} - k^2; \tag{3.29}$$

therefore, in general, the connection of the logarithmic derivatives, i.e., the existence
of a solution valid in both regions, will be possible only for certain energy values.
Since, as we have noted, for large values of ℓ we do not have eigenvalues, the equation
(3.28) must be resolved starting from $\ell = 0, 1, 2, \ldots$ until (3.23) is satisfied.

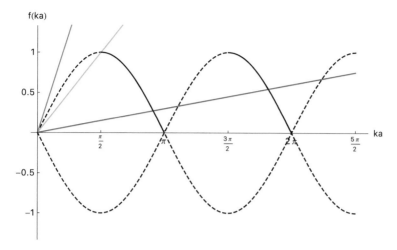

Fig. 3.2 Spherical well: graphic solution to Eq. (3.31). The functions at the two sides of (3.31) are quoted. The left side must be considered only with regard to the continuous black lines. The linear function on the right side is plotted for 3 different values of the angular coefficient: $0.3/\pi$ (red), $2/\pi$ (green), $5/\pi$ (blue)

Eigenvalues for $\ell = 0$

The lowest eigenvalue, the ground state energy, will be obtained for $\ell = 0$: in fact, increasing the orbital quantum number, the well bottom is raised due to the centrifugal potential, and the eigenvalues will consequently be higher. From (A.46) and (A.58), we see that

$$j_0(z) = \frac{\sin z}{z} \quad \text{and} \quad h_0^{(1)}(z) = -\imath \frac{e^{\imath z}}{z}.$$

Substituting them in (3.28), we find that

$$\chi = -k \cot ka. \tag{3.30}$$

This is exactly the same equation already found in the case of the one dimensional square well (problem 2.7) for the odd solutions. This is an expected result, because, for $\ell = 0$, the centrifugal potential is absent. The only difference is linked to the solutions' domain that is restricted to the positive semi-axis with the condition that the radial wave function vanishes in the origin: in the case of the one-dimensional square well, this happens precisely for the odd solutions. However, we are going to solve (3.30), always in a graphic way, but using a different strategy (Fig. 3.2).

Using (3.29) and (3.30), we get

$$\sin ka = \pm\sqrt{\frac{1}{1 + \cot^2 ka}} = \pm\sqrt{\frac{1}{1 + \frac{\chi^2}{k^2}}} = \pm\sqrt{\frac{1}{1 + \frac{\frac{2mV_0}{\hbar^2} - k^2}{k^2}}} = \pm\sqrt{\frac{\hbar^2 k^2}{2mV_0}},$$

which we can also rewrite in the form

$$\pm \sin ka = \sqrt{\frac{\hbar^2}{2mV_0a^2}} \, ka. \tag{3.31}$$

Energy eigenvalues will be obtained substituting the values of ka for which the first and second side curves intersect in (3.29). We must consider that the second side is always positive, and also that ka is positive. Moreover, we remember that equation (3.30) tells us that the cotangent must be negative. It is therefore necessary to consider only the intersections with the continuous part of the *sine* curves in Fig. (3.2). It is evident that there are no solutions if the angular coefficient of the straight line has values too high: it must not be higher than $2/\pi$, that is,

$$\sqrt{\frac{\hbar^2}{2mV_0a^2}} \leq \frac{2}{\pi} \quad \Leftrightarrow \quad V_0a^2 \geq \frac{\pi^2\hbar^2}{8m}.$$

We conclude by noting that, unlike the one-dimensional case, in which there is always at least one positive parity state, in three dimensions, to tie a particle, the well must be deep enough and wide enough, or, better yet, the product V_0a^2 must be large enough. This can be understood by remembering that the wave function must go to zero in the origin and for large r. The connection between these two trends cannot occur if the curvature (the second derivative), which, as seen from (3.24), increases linearly with V_0, is not large enough or if the well is not wide enough.

3.15 Particle in a Nucleus

A nucleus $5 \ 10^{-13}$ cm wide is schematized as a potential well 10 MeV deep. Find the minimum mass for a particle to be inside of the nucleus.

Solution

Remember that (see problem 3.14) the bound states with angular momentum $\ell = 0$ of a spherical well having radius a and depth V_0 are obtained solving Eq. (3.31):

$$\sin ka = \pm \sqrt{\frac{\hbar^2}{2mV_0a^2}} \, ka \quad (\cot ka < 0),$$

where $k = \sqrt{2m(E + V_0)/\hbar^2}$.

   This equation has solutions only if

$$\sqrt{\frac{\hbar^2}{2mV_0a^2}} < \frac{2}{\pi}$$

$$mc^2 > \frac{\pi^2 \hbar^2 c^2}{8 V_0 a^2}.$$

In the present case, posing that $V_0 = 10$ MeV, $a = 5 \ 10^{-13}$ cm, one obtains

$$mc^2 > 192.3 \text{ MeV}.$$

One could ask what happens for higher angular momentum states. We know that, to get states with $\ell > 0$, at fixed mass m, we need higher values of V_0, because the effect of the centrifugal potential is to lift the bottom of the well. Since physical results depend on the product $m V_0$, we can deduce that, if you leave V_0 fixed, the existence of such states requires higher mass values than found for $\ell = 0$. Thus, the mass value found above is indeed a minimum value.

3.16 Particle in a Central Potential

A particle in a potential $V(r)$ is in a state described by the wave function

$$\psi_E(r, \vartheta, \varphi) = A e^{-\frac{r}{a_0}} \qquad (a_0 \text{ constant}),$$

which is an eigenfunction of the Hamiltonian.

(a) What is the angular momentum content of this state?
(b) Assuming that the potential vanishes in the limit $r \to \infty$, find the energy eigenvalue E, considering, in this limit the radial Schrödinger equation

$$\left\{ -\frac{\hbar^2}{2m} \left[\frac{1}{r} \frac{\partial^2}{\partial r^2} r - \frac{\ell(\ell+1)}{r^2} \right] + V(r) \right\} \psi_E(r, \vartheta, \varphi) = E \psi_E(r, \vartheta, \varphi).$$

(c) From the value of E derive $V(r)$, again using the radial equation.

Solution

(a) The wave function does not depend on ϑ and φ, thus the system is in a state with $\ell = m = 0$.
(b) Since

$$\frac{1}{r} \frac{\partial^2}{\partial r^2} r e^{-\frac{r}{a_0}} = \frac{1}{r} \frac{\partial}{\partial r} \left(1 - \frac{r}{a_0} \right) e^{-\frac{r}{a_0}} = \frac{1}{r} \left(-\frac{1}{a_0} - \frac{1}{a_0} + \frac{r}{a_0^2} \right) e^{-\frac{r}{a_0}} =$$

$$= \left[-\frac{2}{a_0} \frac{1}{r} + \frac{1}{a_0^2} \right] e^{-\frac{r}{a_0}},$$

substituting the wave function in the radial equation, we obtain

$$\frac{\hbar^2}{2m}\frac{2}{a_0}\frac{1}{r} - \frac{\hbar^2}{2m}\frac{1}{a_0^2} + V(r) = E.$$

In the limit $r \to \infty$, we find the energy eigenvalue:

$$E = -\frac{\hbar^2}{2ma_0^2}.$$

(c) By replacing this value of E in the previous equation, we have

$$V(r) = -\frac{\hbar^2}{ma_0}\frac{1}{r}.$$

If a_0 is the Bohr radius ($a_0 = \frac{\hbar^2}{me^2}$), we get the potential of the Hydrogen atom $V(r) = -\frac{e^2}{r}$.

3.17   Charged Particle in a Magnetic Field

The Hamiltonian for a charged spin-free particle in a magnetic field $\mathbf{B} = \nabla \times \mathbf{A}$ is

$$\mathcal{H} = \frac{1}{2m}\left(\mathbf{p} - \frac{e}{c}\mathbf{A}(\mathbf{r})\right)^2,$$

where e is the particle's charge and $\mathbf{p} = (p_x, p_y, p_z)$ is the momentum conjugated to the position $\mathbf{r}$. Given that $\mathbf{A} = (-B_0 y, 0, 0)$, corresponding to a constant magnetic field $\mathbf{B} = (0, 0, B_0)$:

(a) Prove that p_x and p_z are constants of motion.
(b) Find the energy levels for this system.

Solution

(a) The system's Hamiltonian can be written in terms of momentum components as:

$$\mathcal{H} = \frac{1}{2m}\left[\left(p_x + \frac{eB_0}{c}y\right)^2 + p_y^2 + p_z^2\right].$$

Therefore, it commutes with all of the $\mathbf{p}$ components except p_y. Remembering Heisemberg's equation, we infer that p_x and p_z are conserved, because they do not depend explicitly on time.

(b) So, we can choose, as a complete set of commuting variables, $\{\mathcal{H}, p_x, p_z\}$. If we call p_x and p_z the eigenvalues of the last two operators, the eigenfunctions have the form

$$\psi(x, y, z) = e^{i\frac{xp_x + zp_z}{\hbar}}\phi(y),$$

where $\phi(y)$ must satisfy the eigenvalue equation

$$\frac{1}{2m}\left[\left(p_x + \frac{eB_0}{c}y\right)^2 - \hbar^2\frac{d^2}{dy^2} + p_z^2\right]\phi(y) = E\phi(y),$$

that is,

$$\left[-\frac{\hbar^2}{2m}\frac{d^2}{dy^2} + \frac{1}{2}m\left(\frac{eB_0}{mc}\right)^2\left(y + \frac{cp_x}{eB_0}\right)^2\right]\phi(y) = \left(E - \frac{p_z^2}{2m}\right)\phi(y).$$

Setting

$$\omega = \frac{|e|B_0}{mc}, \qquad y' = y + \frac{cp_x}{eB_0}, \qquad E' = E - \frac{p_z^2}{2m},$$

we find the equation

$$-\frac{\hbar^2}{2m}\frac{d^2\phi(y)}{dy^2} + \frac{1}{2}m\omega^2 y'^2\phi(y) = E'\phi(y),$$

which is the energy eigenvalue equation for a particle in a one-dimensional harmonic potential with angular frequency ω. The eigenvalues are

$$E' = E - \frac{p_z^2}{2m} = \left(n + \frac{1}{2}\right)\hbar\omega \qquad \text{with} \quad n = 0, 1, 2, \ldots$$

The system's energy eigenvalues are, therefore, given by

$$E_n = \frac{p_z^2}{2m} + \left(n + \frac{1}{2}\right)\hbar\omega \qquad \text{with} \quad n = 0, 1, 2, \ldots$$

3.18 Bound States of the Hydrogenlike Atom

A Hydrogenlike atom is an atom in which there is a positive charge nucleus $+Ze$ and a negative charge electron $-e$. In the case of the Hydrogen atom, we have $Z = 1$, while, for Z different, we have ionized atoms with a single electron.

The Coulomb potential energy is, in Gauss units,

$$V(r) = -\frac{Ze^2}{r}. \tag{3.32}$$

The atomic states are the bound states ($E < 0$) for this potential.

We introduce the quantities χ related to the eigenvalue E

$$\chi = \sqrt{-\frac{2mE}{\hbar^2}} \tag{3.33}$$

and

$$\lambda = \frac{Ze^2 m}{\chi \hbar^2} = z\alpha \sqrt{\frac{mc^2}{2|E|}}, \tag{3.34}$$

where

$$\alpha = \frac{e^2}{\hbar c} \simeq \frac{1}{137} \tag{3.35}$$

is the fine-structure constant.

In terms of these parameters, the radial equation (A.43) becomes

$$\frac{d^2}{dr^2} U_{E,\ell}(r) + \left[-\chi^2 + \frac{2\chi\lambda}{r} - \frac{\ell(\ell+1)}{r^2} \right] U_{E,\ell}(r) = 0. \tag{3.36}$$

Calculate, solving this equation by the power expansion method, the energy eigenvalues corresponding to bound states.

Solution

We note that the Coulomb potential is less divergent than r^{-2} in the origin. We know that, in these cases, the only regular solution in $r = 0$ must have the behavior

$$U_{E,\ell}(r) \underset{r \to 0}{\sim} r^{\ell+1}, \tag{3.37}$$

while, as we have seen in discussing the spherical well spectrum (problem 3.14), since, in the limit $r \to \infty$, the potential goes to zero, it must have the same behavior as the spherical functions of Hankel of the first kind, that is,

$$U_{E,\ell}(r) \underset{r \to \infty}{\sim} e^{-\chi r}. \tag{3.38}$$

These behaviors suggest the factorization

$$U_{E,\ell}(r) = r^{\ell+1} e^{-\chi r} u(r), \tag{3.39}$$

where $u(r)$ is an interpolating function to be determined. Substituting (3.39) and introducing the dimensionless variable

$$t = 2\chi r, \tag{3.40}$$

the radial equation 3.36 becomes an equation for the interpolating function $u(t)$:

$$t \frac{d^2}{dt^2} u(t) + (2\ell + 2 - t) \frac{d}{dt} u(t) - (\ell + 1 - \lambda) u(t) = 0. \tag{3.41}$$

This equation is the Kummer confluent hypergeometric equation. We search for solutions satisfying the behaviors in (3.37) and (3.38) through the power expansion method. Substituting the expansion

$$u(t) = \sum_{k=0}^{\infty} a_k t^k \tag{3.42}$$

in (3.41), we find

$$\sum_{k=2}^{\infty} k(k-1)a_k t^{k-1} + (2\ell + 2 - t)\sum_{k=1}^{\infty} k a_k t^{k-1} - (\ell + 1 - \lambda)\sum_{k=0}^{\infty} a_k t^k = 0.$$

We redefine the sum index $k - 1 \to k$ in the first two summations and separate the second summation into two terms:

$$\sum_{k=1}^{\infty}(k+1)k a_{k+1} t^k + (2\ell + 2)\sum_{k=0}^{\infty}(k+1)a_{k+1} t^k - \sum_{k=1}^{\infty} k a_k t^k - (\ell + 1 - \lambda)\sum_{k=0}^{\infty} a_k t^k = 0.$$

Noting that, in the first and third terms, the sum index can start from 0, we are able to aggregate the terms that have the same coefficients a_k:

$$\sum_{k=0}^{\infty}\left[(k+1)(k+2\ell+2)\,a_{k+1} - (\ell + 1 - \lambda + k)\,a_k\right]t^k = 0.$$

Imposing term-by-term vanishing we find the following recursion relation:

$$\frac{a_{k+1}}{a_k} = \frac{\ell + 1 - \lambda + k}{(k+1)(k+2\ell+2)}. \tag{3.43}$$

This ratio behaves, in the limit $k \to \infty$, as

$$\frac{a_{k+1}}{a_k} \underset{k\to\infty}{\sim} \frac{1}{k}.$$

It follows that the power series (3.42) has the same behavior for large r of $e^t = \sum_{k=0}^{\infty}\frac{t^k}{k!}$. If this occurs, the asymptotic trend of the radial function would be

$$U_{E,\ell}(r) = r^{\ell+1}\,e^{-\chi r}\sum_{k=0}^{\infty} a_k (2\chi r)^k \underset{r\to\infty}{\sim} e^{\chi r},$$

which is not acceptable. However, this would not happen if the series were truncated starting from a certain k, i.e., if there exists a n_r value of k such that

$$\lambda = n_r + \ell + 1 \quad \text{with} \quad n_r = 0, 1, 2, \dots$$

n_r, called the *radial quantum number*, is the degree of the polynomial to which the series is reduced. Note that λ, being the sum of integers, is also an integer, and we can pose $\lambda = n$, so that

$$n = n_r + \ell + 1 \quad \text{with} \quad n = 1, 2, \dots \tag{3.44}$$

n is called the *principal quantum number*. The Eq. (3.34) allows us to determine the energy spectrum:

$$E_n = -\frac{1}{2} mc^2 \frac{(Z\alpha)^2}{n^2}. \tag{3.45}$$

Equation (3.44) shows that these eigenvalues are degenerate.

3.19 Expectation Values of $\frac{1}{r^n}$ for $n = 1, 2, 3$ in the Hydrogenlike Atom Stationary States

Using the Feynman-Hellmann theorem (see problem 1.12)

$$\left\langle \frac{\partial \mathcal{H}(\lambda)}{\partial \lambda} \right\rangle = \frac{\partial E(\lambda)}{\partial \lambda}, \tag{3.46}$$

calculate the expectation values of $\frac{1}{r^n}$ for $n = 1, 2, 3$ in the stationary states of the Hydrogenlike atom.

Solution

(a) **Calculation of** $\left\langle \frac{1}{r} \right\rangle_{n,\ell}$

We use, as a parameter, the fine-structure constant α, which, in the Hydrogenlike atom Hamiltonian, is only present in the Coulomb potential energy:

$$V_{coul}(r) = -\frac{e^2}{r} = -\frac{\hbar c \alpha}{r}.$$

Applying the Feynman-Hellmann theorem, we have

$$\left\langle \frac{\partial \mathcal{H}(\alpha)}{\partial \alpha} \right\rangle_{n,\ell} = \left\langle -\frac{\hbar c}{r} \right\rangle_{n,\ell} = \frac{\partial E_n(\alpha)}{\partial \alpha} = -\frac{mc^2 \alpha}{n^2},$$

from which we obtain

$$\left\langle \frac{1}{r} \right\rangle_{n,\ell} = \frac{mc\alpha}{\hbar n^2} = -2\frac{E_n}{\hbar c \alpha} = -\frac{2E_n}{e^2}$$

and, finally, the result (which will be used to calculate fine-structure corrections in problem 6.25)

$$\left(\frac{e^2}{r}\right)_{n,\ell} = -2E_n. \tag{3.47}$$

(b) **Calculation of** $\left(\frac{1}{r^2}\right)_{n,\ell}$

This time, for our parameter, we use the orbital quantum number ℓ, which, in the Hydrogenlike atom Hamiltonian, is only present in the centrifugal potential term:

$$V_\ell(r) = \frac{\hbar^2 \ell(\ell + 1)}{2mr^2}.$$

From the Feynman-Hellmann theorem, we obtain

$$\left(\frac{\partial \mathcal{H}(\ell)}{\partial \ell}\right)_{n,\ell} = \frac{\hbar^2}{2m} \left(\frac{2\ell + 1}{r^2}\right)_{n,\ell} =$$

$$= \frac{\partial E_n(\ell)}{\partial \ell} = \frac{\partial}{\partial \ell}\left(-\frac{1}{2} mc^2\alpha^2 \frac{1}{(n_r + \ell + 1)^2}\right) = \frac{1}{2} mc^2\alpha^2 \frac{2}{n^3},$$

from which we get the result (also used to calculate fine-structure corrections)

$$\left(\frac{e^4}{r^2}\right)_{n,\ell} = 8(E_n)^2 \frac{n}{2\ell + 1}. \tag{3.48}$$

(c) **Calculation of** $\left(\frac{1}{r^3}\right)_{n,\ell}$

This calculation derives from the previous result, thanks to the fact that, in a stationary state, the average force must be zero. To realize this, just write the impulse expectation value in this state; one immediately understands that it is not modified by the temporal evolution and, therefore, its derivative with respect to time, which is precisely the average force, must be zero. In the present case, the force is

$$F(r) = -\frac{dV(r)}{dr} = -\frac{d}{dr}\left(-\frac{e^2}{r} + \frac{\hbar^2 \ell(\ell + 1)}{2mr^2}\right) = -\frac{e^2}{r^2} + \frac{\hbar^2 \ell(\ell + 1)}{mr^3}.$$

Imposing that $\langle F(r) \rangle_{n,\ell} = 0$ and using the expectation value of $\frac{1}{r^2}$ (3.48), we obtain

$$\left(\frac{1}{r^3}\right)_{n,\ell} = \left(\frac{e^2}{r^2}\right)_{n,\ell} \frac{m}{\hbar^2 \ell(\ell + 1)} = \frac{1}{a_0^3} \frac{1}{n^3 \ell(\ell + \frac{1}{2})(\ell + 1)} \tag{3.49}$$

(result to be used to calculate corrections due to spin-orbit interaction in problem 6.26).

3.20 One-Dimensional Hydrogen Atom? A Misleading Similarity

The state of a particle of mass m in one dimension is described by the wave function

$$\psi(x) = A \left(\frac{x}{x_0} \right)^n e^{-\frac{x}{x_0}},$$

where A, n e x_0 are constants.

(a) Using the Schrödinger equation, find the potential $V(x)$ and the energy eigenvalue E for which this wave function is an energy eigenfunction (assume that $V(x) \to 0$ when $x \to \infty$).

(b) What connection can be noted between this potential and the effective (coulombian + centrifugal) radial potential for a Hydrogen atom in the orbital angular momentum ℓ?

Solution

(a) Replacing $\psi(x)$ in the Schrödinger equation, we obtain

$$[E - V(x)]\psi(x) = -\frac{\hbar^2}{2m} \left[\frac{n(n-1)}{x^2} - 2\frac{n}{xx_0} + \frac{1}{x_0^2} \right] \psi(x).$$

Assuming that $V(x) \xrightarrow[x \to \infty]{} 0$, we get

$$E = -\frac{\hbar^2}{2m} \frac{1}{x_0^2}, \quad \text{and therefore} \quad V(x) = \frac{\hbar^2}{2m} \left[\frac{n(n-1)}{x^2} - 2\frac{n}{xx_0} \right].$$

(b) The term $\frac{\hbar^2}{2m} \frac{n(n-1)}{x^2}$ is similar to the centrifugal potential $\frac{\hbar^2}{2m} \frac{\ell(\ell+1)}{r^2}$ in the radial equation, but the term that depends on $\frac{1}{x}$ contains n, which obviously does not happen for the Coulomb potential $\frac{e^2}{r}$.

3.21 Determining the State of a Hydrogen Atom

Of a Hydrogen atom, it is known that:

(a) it is in a p state with $n=2$,
(b) it contains eigenstates of L_z corresponding to eigenvalues $+1$ and -1,
(c) the expectation value of L_z is zero,
(d) the probability of finding the electron in the first quadrant ($0 < \phi < \frac{\pi}{2}$) is 25%.

Write down the possible wave functions.

Solution

We denote by $|n\ell m\rangle$ the generic eigenstate common to $\mathcal{H}, L^2, L_z$ of a Hydrogen atom. Conditions (a) and (b) allow us to write the desired state as:

$$|\psi\rangle = \alpha|2, 1, 1\rangle + \beta|2, 1, -1\rangle.$$

From condition (c)

$$\langle\psi|L_z|\psi\rangle = |\alpha|^2\hbar - |\beta|^2\hbar = 0 \quad \Rightarrow \quad |\alpha|^2 = |\beta|^2.$$

By imposing normalization and taking into account that, since the overall phase is undetermined, we can fix α to be real and positive, we get

$$\alpha = |\beta| = \frac{1}{\sqrt{2}}$$

and, having called δ the phase of β,

$$\beta = \frac{1}{\sqrt{2}} e^{i\delta}.$$

To apply condition (d), we need to transition to wave functions in the coordinate basis. The probability of finding the electron between ϕ and $\phi + d\phi$ is obtained by integrating the square modulus of $\psi(r, \theta, \phi) = \langle\mathbf{r}|\psi\rangle$ over the other variables:

$$P(\phi)\,d\phi = \left|\frac{1}{\sqrt{2\pi}}\left(\frac{1}{\sqrt{2}}e^{i\phi} + \frac{1}{\sqrt{2}}e^{-i\phi+i\delta}\right)\right|^2 d\phi.$$

$$P\left(0 < \phi < \frac{\pi}{2}\right) = \int_0^{\frac{\pi}{2}} P(\phi)d\phi =$$

$$= \frac{1}{2\pi} \int_0^{\frac{\pi}{2}} [1 + \cos(2\phi - \delta)]\,d\phi =$$

$$= \frac{1}{4} + \frac{1}{2\pi}\sin\delta = \frac{1}{4}. \tag{3.50}$$

Then,

$$\sin\delta = 0 \quad \Rightarrow \quad \delta = n\pi.$$

So, we have two possible status determinations corresponding to the choice of n even or odd:

$$|\psi\rangle = \begin{cases} \frac{1}{\sqrt{2}}|2, 1, 1\rangle + \frac{1}{\sqrt{2}}|2, 1, -1\rangle \\[2mm] \frac{1}{\sqrt{2}}|2, 1, 1\rangle - \frac{1}{\sqrt{2}}|2, 1, -1\rangle. \end{cases}$$

3.22 Hydrogen Atom in the Ground State

The wave function of the ground state of the Hydrogen atom is

$$\psi_{1,0,0} = \sqrt{\frac{1}{\pi a_0^3}}\, e^{-\frac{r}{a_0}},$$

where $a_0 = \hbar^2/me^2$ is the Bohr radius.

(a) Determine the distance from the nucleus at which the probability density of finding the electron is maximum.

(b) Also determine the expectation value of the electron's position.

Solution

We need the probability of finding the electron at a fixed distance from the nucleus (or, rather, to find the reduced mass at a fixed distance from the center of mass) independently of its direction, so the probability distribution must be integrated over the whole solid angle:

$$P(r)\,dr = \int d\Omega\, |\psi_{1,0,0}(r,\theta,\phi)|^2\, r^2\,dr = 4\pi\,|R_{1,0}(r)|^2 r^2 dr = \frac{4r^2}{a_0^3}\, e^{-\frac{2r}{a_0}}\, dr,$$

where $R_{1,0}(r)$ is the radial function (A.42) of the Hydrogen ground state.

(a) This probability density is maximum when r is the solution to the equation

$$\frac{dP(r)}{dr} = \frac{4}{a_0^3}\left[2r - 2\frac{r^2}{a_0}\right] e^{-\frac{2r}{a_0}} = 0 \quad \text{with} \quad \frac{d^2P(r)}{dr^2} < 0.$$

Therefore, the required maximum is at $r = a_0$ ($r = 0$ corresponds to a minimum).

(b) We get the expectation value of the electron's position using the formula (A.6):

$$<r> = \int_0^\infty dr\, r\, P(r) = \frac{4}{a_0^3}\int_0^\infty dr\, r^3\, e^{-\frac{2r}{a_0}} = \frac{a_0}{4}\int_0^\infty d\alpha\, \alpha^3\, e^{-\alpha} = \frac{3}{2}\, a_0.$$

3.23 Hydrogen Atom in an External Magnetic Field

Consider a Hydrogen atom in the state $2p$ (neglecting the electron spin) and use the basis common to operators $\mathcal{H}, L^2, L_z$.

(a) Denote by $|\psi_+\rangle, |\psi_0\rangle, |\psi_-\rangle$ the normalized state vectors corresponding to $m = +1, 0, -1$, respectively. Immerse the Hydrogen atom in an external magnetic field $\mathbf{B}$ parallel to the z axis and let the interaction energy be given by

$$W = -\beta \, \mathbf{B} \cdot \mathbf{L}.$$

Determine the new system energy levels of the states $2p$.

(b) Consider the state

$$|\psi\rangle = \frac{1}{2}(|\psi_+\rangle + \sqrt{2}|\psi_0\rangle + |\psi_-\rangle).$$

Calculate $< E >$ and $\Delta E^2 = < (E - < E >)^2 >$ in this state.

(c) Limiting ourselves to the subspace $2p$, L_z is represented by a diagonal matrix in the basis $|\psi_+\rangle$, $|\psi_0\rangle$, $|\psi_-\rangle$, while

$$L_x = \frac{\hbar}{\sqrt{2}} \begin{pmatrix} 0 & 1 & 0 \\ 1 & 0 & 1 \\ 0 & 1 & 0 \end{pmatrix},$$

$$L_y = \frac{\hbar}{i\sqrt{2}} \begin{pmatrix} 0 & 1 & 0 \\ -1 & 0 & 1 \\ 0 & -1 & 0 \end{pmatrix}.$$

Calculate the L_x and L_y expectation values in the state $|\psi\rangle$.

Solution

(a) The energy of the $2p$ levels in the absence of the magnetic field is

$$E_2 = -\frac{\mu c^2 \alpha^2}{8},$$

where μ is the electron's reduced mass and α is the fine structure constant. The energy contribution due to the magnetic field corresponds to the operator

$$W = -\beta B L_z,$$

which commutes with the remaining part of the Hamiltonian. Moreover, the $2p$ states are eigenstates of L_z, so the new energy levels are

$$E_2^{+1} = -\frac{\mu c^2 \alpha^2}{8} - \beta \hbar B, \qquad E_2^0 = -\frac{\mu c^2 \alpha^2}{8}, \qquad E_2^{-1} = -\frac{\mu c^2 \alpha^2}{8} + \beta \hbar B,$$

where the levels were indexed with the value of the quantum number m. The presence of the external magnetic field, therefore, generated a breakdown of the degeneration.

(b) The state vector $|\psi\rangle$ is normalized. So, we have

$$\langle E \rangle = \left(\frac{1}{2}\right)^2 \left[1^2 E_2^{+1} + (\sqrt{2})^2 E_2^0 + 1^2 E_2^{-1}\right] = -\frac{\mu c^2 \alpha^2}{8},$$

$$\langle E^2 \rangle = \frac{1}{4}\left[(E_2^{+1})^2 + 2(E_2^0)^2 + (E_2^{-1})^2\right] = \left(\frac{\mu c^2 \alpha^2}{8}\right)^2 + \frac{1}{2}(\beta \hbar B)^2,$$

$$\Delta E^2 = \langle (E - \langle E \rangle)^2 \rangle = \langle E^2 \rangle - \langle E \rangle^2 = \frac{1}{2}\beta^2 \hbar^2 B^2.$$

(c) In the suggested basis, the state vector $|\psi\rangle$ corresponds to the one column matrix

$$|\psi\rangle = \frac{1}{2}\begin{pmatrix} 1 \\ \sqrt{2} \\ 1 \end{pmatrix}.$$

Therefore, we get

$$< L_x >= \frac{1}{2}\left(1\ \sqrt{2}\ 1\right)\frac{\hbar}{\sqrt{2}}\begin{pmatrix} 0 & 1 & 0 \\ 1 & 0 & 1 \\ 0 & 1 & 0 \end{pmatrix}\frac{1}{2}\begin{pmatrix} 1 \\ \sqrt{2} \\ 1 \end{pmatrix} = \hbar,$$

$$< L_y >= \frac{1}{2}\left(1\ \sqrt{2}\ 1\right)\frac{\hbar}{i\sqrt{2}}\begin{pmatrix} 0 & 1 & 0 \\ -1 & 0 & 1 \\ 0 & -1 & 0 \end{pmatrix}\frac{1}{2}\begin{pmatrix} 1 \\ \sqrt{2} \\ 1 \end{pmatrix} = 0.$$

3.24 A Molecular Model

For some molecules, the potential energy can be modeled with the expression

$$V(r) = -2D\left(\frac{a}{r} - \frac{a^2}{2r^2}\right).$$

Determine the energy levels for this potential energy and discuss these results in the hypothesis, often valid, $D \gg \frac{\hbar^2}{2ma^2}$.

Solution

Introducing the variable

$$\rho = \frac{r}{a}$$

and the parameters

$$\kappa^2 = -\frac{2ma^2}{\hbar^2}E \quad \text{and} \quad \gamma^2 = \frac{2ma^2}{\hbar^2}D,$$

the radial Schrödinger equation (A.43) becomes

$$U_{\ell''}(\rho) + \left[-\kappa^2 + \frac{2\gamma^2}{\rho} - \frac{\gamma^2 + \ell(\ell+1)}{\rho^2}\right] U_{\ell}(\rho) = 0.$$

This equation is similar to that for the Hydrogen atom (problem 3.18)

$$U_{\ell''}(r) + \left[-\kappa^2 + \frac{2\epsilon\lambda}{r} - \frac{\ell'(\ell'+1)}{r^2}\right] U_{\ell}(r) = 0,$$

except for the substitutions

$$\lambda = \frac{\gamma^2}{\epsilon} \quad \text{and} \quad \gamma^2 + \ell(\ell+1) = \ell'(\ell'+1) \Leftrightarrow \ell' = \sqrt{\gamma^2 + \left(\ell + \frac{1}{2}\right)^2} - \frac{1}{2}.$$

In the case of the Hydrogen atom, the requirement that the wave function be regular at infinity leads to the quantization condition

$$\ell' + 1 - \lambda = -n_r.$$

In the present case, we get

$$\sqrt{\gamma^2 + (\ell + \frac{1}{2})^2} + \frac{1}{2} - \frac{\gamma^2}{\epsilon} = -n_r.$$

Therefore the energy levels of the bound states are given by

$$E_{n_r,\ell} = -\frac{\hbar^2}{2ma^2}\kappa^2 = -\frac{\hbar^2}{2ma^2}\frac{\gamma^4}{\left[\sqrt{\gamma^2 + (\ell + \frac{1}{2})^2} + \frac{1}{2} + n_r\right]^2} =$$

$$= -D\frac{1}{\left[\sqrt{1 + (\ell + \frac{1}{2})^2 x^2} + (\frac{1}{2} + n_r)x\right]^2},$$

where

$$x = \frac{1}{\gamma}.$$

The $D \gg \frac{\hbar^2}{2ma^2}$ regime corresponds to $\gamma \gg 1$, i.e., $x \ll 1$. We can expand the expression for energy levels in power series of x up to the 2nd order. Taking into account the following series expansions,

$$\sqrt{1 + x^2} \approx 1 + \frac{x^2}{2}$$

$$\left[\frac{1}{1 + ax + bx^2}\right]^2 \approx \left[1 - ax + (a^2 - b)x^2\right]^2 \approx 1 - 2ax + (3a^2 - 2b)x^2,$$

we get

$$E_{n_r,\ell} \approx -D\left[1 - \frac{2(n_r + \frac{1}{2})}{\gamma} - \frac{(\ell + \frac{1}{2})^2}{\gamma^2} + 3\frac{(n_r + \frac{1}{2})^2}{\gamma^2}\right].$$

It is clear that this approximation makes sense only for small values of quantum numbers, otherwise the terms ignored become important because they depend on powers of n_r and ℓ. The three terms can be interpreted:

(a) the first term is a constant related to the value of the potential minimum. Indeed, V at the point of minimum $r = a$ gets the value $-D$;
(b) the second term is a vibrational term with frequency

$$\omega = \sqrt{\frac{2D}{ma^2}},$$

due to the fact that, around the minimum, the potential is approximable with a harmonic oscillator potential;
(c) the third and fourth terms represent rotational energy proportional to

$$\frac{D}{\gamma^2} = \frac{\hbar^2}{2ma^2} = \frac{\hbar^2}{2I}, \quad \text{where } I \text{ is the system moment of inertia.}$$

Chapter 4
Spin

4.1 Total Spin of Two Electrons

Consider a system of two electrons that are in a state with opposite spin z-components. Calculate the S^2 (where $\mathbf{S}$ is the total spin operator) expectation value in this state.

Solution

Suppose the system state is

$$|S_{1,z}, S_{2,z}\rangle = |+, -\rangle.$$

This state contributes to two different eigenstates of the total spin operator:

$$|s = 0, m = 0\rangle = \frac{1}{\sqrt{2}} (|+, -\rangle - |-, +\rangle)$$

and

$$|s = 1, m = 0\rangle = \frac{1}{\sqrt{2}} (|+, -\rangle + |-, +\rangle),$$

where s and m are the quantum numbers of S^2 and S_z. It turns out, therefore, that

$$|+, -\rangle = \frac{1}{\sqrt{2}} (|s = 0, m = 0\rangle + |s = 1, m = 0\rangle).$$

The S^2 expectation value is obtained averaging the eigenvalues of the two eigenstates weighed with the square modulus of their coefficients:

$$\langle S^2 \rangle = \left| \frac{1}{\sqrt{2}} \right|^2 0(0 + 1)\hbar^2 + \left| \frac{1}{\sqrt{2}} \right|^2 1(1 + 1)\hbar^2 = \hbar^2.$$

© Springer Nature Switzerland AG 2019
L. Angelini, *Solved Problems in Quantum Mechanics*, UNITEXT for Physics,
https://doi.org/10.1007/978-3-030-18404-9_4

4.2 Eigenstates of a Spin Component (I)

The spin function of a spin $\frac{1}{2}$ particle has the following expression in the S_z basis:

$$\begin{pmatrix} \psi_1 \\ \psi_2 \end{pmatrix} = \begin{pmatrix} e^{i\alpha} \cos \delta \\ e^{i\beta} \sin \delta \end{pmatrix}.$$

Does an $\hat{n}$ space-direction exist such that the result of the measurement of the spin component along $\hat{n}$ can be predicted with certainty?

Solution

The state is already normalized. Having named as ϑ and φ the spherical angular coordinates of the direction $\hat{n}$, this state must be an eigenstate of the matrix representative of $\mathbf{S} \cdot \hat{n}$ in the basis common to S^2 and S_z:

$$\mathbf{S} \cdot \hat{n} = \frac{\hbar}{2} \left[\sin \vartheta \, \cos \varphi \begin{pmatrix} 0 & 1 \\ 1 & 0 \end{pmatrix} + \sin \vartheta \, \sin \varphi \begin{pmatrix} 0 & -i \\ i & 0 \end{pmatrix} + \cos \vartheta \begin{pmatrix} 1 & 0 \\ 0 & -1 \end{pmatrix} \right] =$$

$$= \frac{\hbar}{2} \begin{pmatrix} \cos \vartheta & \sin \vartheta \, e^{-i\varphi} \\ \sin \vartheta \, e^{i\varphi} & -\cos \vartheta \end{pmatrix}.$$

The eigenvalues of $\mathbf{S} \cdot \hat{n}$, because of the isotropy of space, are the same as those of S_z, i.e., $\pm \frac{\hbar}{2}$, whereas their eigenvectors are:

$$|\mathbf{S} \cdot \hat{n} = +\frac{\hbar}{2}\rangle = \begin{pmatrix} \cos \frac{\vartheta}{2} \\ \sin \frac{\vartheta}{2} e^{i\varphi} \end{pmatrix} \qquad |\mathbf{S} \cdot \hat{n} = -\frac{\hbar}{2}\rangle = \begin{pmatrix} \sin \frac{\vartheta}{2} \\ -\cos \frac{\vartheta}{2} e^{i\varphi} \end{pmatrix}$$

For there to be certainty in the result of a measure, the spin function must be able to identify, less than an arbitrary phase factor, with one of these eigenstates. There are, therefore, two possibilities:

$$\varphi = \beta - \alpha, \ \delta = \frac{\vartheta}{2}$$

or

$$\varphi = \beta - \alpha, \ \delta = \frac{\vartheta + \pi}{2}.$$

4.3 Eigenstates of a Spin Component (II)

If a particle of spin $\frac{1}{2}$ is in a state with $S_z = \frac{\hbar}{2}$, what are the probabilities that, measuring the component of the spin along the direction $\hat{n}$ identified by the angular spherical coordinates ϑ and φ, one will find $\pm \frac{\hbar}{2}$?

Solution

As we have seen in the problem 4.2, the eigenstates of $\mathbf{S} \cdot \hat{n}$ are

$$|S \cdot \hat{n} = +\frac{\hbar}{2}\rangle = \begin{pmatrix} \cos\frac{\vartheta}{2} \\ \sin\frac{\vartheta}{2} e^{i\varphi} \end{pmatrix} = \cos\frac{\vartheta}{2}\begin{pmatrix} 1 \\ 0 \end{pmatrix} + \sin\frac{\vartheta}{2} e^{i\varphi}\begin{pmatrix} 0 \\ 1 \end{pmatrix},$$

$$|S \cdot \hat{n} = -\frac{\hbar}{2}\rangle = \begin{pmatrix} \sin\frac{\vartheta}{2} \\ -\cos\frac{\vartheta}{2} e^{i\varphi} \end{pmatrix} = \sin\frac{\vartheta}{2}\begin{pmatrix} 1 \\ 0 \end{pmatrix} - \cos\frac{\vartheta}{2} e^{i\varphi}\begin{pmatrix} 0 \\ 1 \end{pmatrix}.$$

In terms of these eigenstates, the particle state before the measure is given by

$$\begin{pmatrix} 1 \\ 0 \end{pmatrix} = \cos\frac{\vartheta}{2}|S \cdot \hat{n} = +\frac{\hbar}{2}\rangle - \sin\frac{\vartheta}{2}|S \cdot \hat{n} = -\frac{\hbar}{2}\rangle.$$

As a result, the required probabilities are

$$P\left(S \cdot \hat{n} = +\frac{\hbar}{2}\right) = \cos^2\frac{\vartheta}{2} \quad \text{and} \quad P\left(S \cdot \hat{n} = -\frac{\hbar}{2}\right) = \sin^2\frac{\vartheta}{2}.$$

4.4 Determining a Spin State (I)

An electron is equally likely to have its spin oriented parallel or antiparallel to the z axis. Determine its status when the expectation value of S_x is maximum.

Solution

In the S_z representation, the electron state is

$$\psi = \begin{pmatrix} a \\ b \end{pmatrix}.$$

By imposing the condition given by the problem, we know that the coefficients a and b have the same modulus. Then, from normalization, and taking into account the arbitrariness of the overall phase, we get

$$a = \frac{1}{\sqrt{2}} \quad \text{and} \quad b = \frac{1}{\sqrt{2}} e^{i\alpha}$$

and

$$\psi = \frac{1}{\sqrt{2}}\begin{pmatrix} 1 \\ e^{i\alpha} \end{pmatrix}.$$

The S_x expectation value is

$$\langle S_x \rangle = \frac{1}{\sqrt{2}} \left(1 \; e^{-i\alpha} \right) \frac{\hbar}{2} \begin{pmatrix} 0 & 1 \\ 1 & 0 \end{pmatrix} \frac{1}{\sqrt{2}} \begin{pmatrix} 1 \\ e^{i\alpha} \end{pmatrix} = \frac{\hbar}{4} \left(1 \; e^{-i\alpha} \right) \begin{pmatrix} e^{i\alpha} \\ 1 \end{pmatrix} =$$

$$= \frac{\hbar}{4} \left(e^{i\alpha} + e^{-i\alpha} \right) = \frac{\hbar}{2} \cos \alpha,$$

which has $\frac{\hbar}{2}$ as its maximum value when $\alpha = 0$ and

$$a = b = \frac{1}{\sqrt{2}}.$$

We note that the maximum value of $\langle S_x \rangle$ found is equal to its maximum eigenvalue; in fact, the state we found is the eigenstate of S_x corresponding to this eigenvalue.

4.5   Determining a Spin State (II)

An electron is in a S_z eigenstates superposition:

$$|\psi\rangle = a \, |+\rangle + b \, |-\rangle.$$

Determine the constants a and b so that the expectation values of S_z and S_y are 0 and $\frac{\hbar}{2}$, respectively.

Solution

In the representation in which S_z is diagonal the state of the electron is given by

$$\psi = \begin{pmatrix} a \\ b \end{pmatrix}.$$

In this state the expectation values of S_z and S_y are

$$\langle S_z \rangle = \frac{\hbar}{2} \left(a^* \; b^* \right) \begin{pmatrix} 1 & 0 \\ 0 & -1 \end{pmatrix} \begin{pmatrix} a \\ b \end{pmatrix} = \frac{\hbar}{2} \left(|a|^2 - |b|^2 \right),$$

$$\langle S_y \rangle = \frac{\hbar}{2} \left(a^* \; b^* \right) \begin{pmatrix} 0 & -i \\ i & 0 \end{pmatrix} \begin{pmatrix} a \\ b \end{pmatrix} = -\frac{\hbar}{2} i (a^* b - a b^*) = -\frac{\hbar}{2} 2 \Im(a b^*).$$

From the condition $\langle S_z \rangle = 0$ and from the normalization relation, we obtain that coefficients a and b have the same modulus and, placing the phase of a as being equal to 0, we get

$$a = \frac{1}{\sqrt{2}} \quad \text{and} \quad b = \frac{1}{\sqrt{2}} e^{i\alpha}.$$

The phase α can be determined from the condition $\langle S_y \rangle = \frac{\hbar}{2}$:

$$-2\Im(ab^*) = 1 \quad \Rightarrow \quad \sin\alpha = 1 \quad \Rightarrow \quad \alpha = \frac{\pi}{2}$$

In conclusion,

$$|\psi\rangle = \frac{1}{\sqrt{2}}(|+\rangle + i\,|-\rangle).$$

4.6 Determining a Spin State (III)

A spin $\frac{1}{2}$ particle is in a state in which the expectation value of S_x is $\frac{\hbar}{2}\alpha$ and that of S_y is $\frac{\hbar}{2}\beta$ with α and β between -1 and 1.

Show that the condition $\alpha^2 + \beta^2 \le 1$ must hold and that, for $\alpha^2 + \beta^2 < 1$, the problem admits two solutions, while, for $\alpha^2 + \beta^2 = 1$, the solution is unique. In the latter case, calculate the probability of finding the particle spin oriented parallel or antiparallel compared to the z axis.

Solution

Having called $|\psi\rangle$ the state under examination and $|\pm\rangle$ the eigenstates of S_z with eigenvalues $\pm\hbar/2$, we have

$$|\psi\rangle = a\,|+\rangle + b\,|-\rangle = \begin{pmatrix} a \\ b \end{pmatrix},$$

where a and b are two constants to be determined. The expectation values of S_x and S_y in this state are

$$\langle S_x \rangle = \frac{\hbar}{2}\begin{pmatrix} a^* & b^* \end{pmatrix}\begin{pmatrix} 0 & 1 \\ 1 & 0 \end{pmatrix}\begin{pmatrix} a \\ b \end{pmatrix} = \frac{\hbar}{2}(a^*b + ab^*),$$

$$\langle S_y \rangle = \frac{\hbar}{2}\begin{pmatrix} a^* & b^* \end{pmatrix}\begin{pmatrix} 0 & -i \\ i & 0 \end{pmatrix}\begin{pmatrix} a \\ b \end{pmatrix} = -\frac{\hbar}{2}i(a^*b - ab^*).$$

We can choose the phases of a and b so that

$$a > 0 \quad b = |b|\,e^{i\vartheta}.$$

Imposing normalization, we get

$$|a|^2 + |b|^2 = 1 \quad \Rightarrow \quad |b|^2 = 1 - a^2 \quad \Rightarrow \quad b = \sqrt{1 - a^2}\,e^{i\vartheta}.$$

The expectation values become

$$\frac{2}{\hbar}\langle S_x \rangle = a\,2\Re(b) = 2a\sqrt{1-a^2}\,\cos\vartheta,$$

$$\frac{2}{\hbar}\langle S_y \rangle = -ia\,2i\Im(b) = 2a\sqrt{1-a^2}\,\sin\vartheta,$$

and, by imposing the conditions given, we obtain

$$\alpha^2 + \beta^2 = \frac{4}{\hbar^2}\left(\langle S_x\rangle^2 + \langle S_y\rangle^2\right) = 4a^2(1-a^2) \quad \text{with}\ \ a^2 \le 1.$$

The right side of this equation is, in the variable a^2, a parabola with its concavity facing downwards, symmetrical with respect to the axis $a^2 = 1/2$, which assumes its maximum value in $a^2 = 1/2$, so that

$$\alpha^2 + \beta^2 \le 4a^2(1-a^2)\big|_{a^2=\frac{1}{2}} = 1.$$

Any other value of $\alpha^2 + \beta^2 < 1$ corresponds to two values of a^2 symmetrical with respect to $a^2 = 1/2$.

In the case $\alpha^2 + \beta^2 = 1$, we have $a = 1/\sqrt{2} = |b|$, and therefore

$$|\psi\rangle = \frac{1}{\sqrt{2}}\,|+\rangle + \frac{1}{\sqrt{2}}\,e^{i\vartheta}\,|-\rangle.$$

So, the required probabilities are both equal to $1/2$.

4.7 Measurements in a Stern-Gerlach Apparatus

A beam of spin $\frac{1}{2}$ atoms moving in the direction of the y axis is subjected to a series of measurements by Stern-Gerlach-type devices as follows:

(a) The first measurement accepts atoms with $s_z = \frac{\hbar}{2}$ and rejects atoms with $s_z = -\frac{\hbar}{2}$.

(b) The second measurement accepts atoms with $s_n = \frac{\hbar}{2}$ and rejects atoms with $s_n = -\frac{\hbar}{2}$, where s_n is eigenvalue of the operator $\mathbf{S}\cdot\hat{\mathbf{n}}$ and $\hat{\mathbf{n}}$ the unit vector in the xz-plane at an angle ϑ with respect to the z-axis.

(c) The third measurement accepts atoms with $s_z = -\frac{\hbar}{2}$ and rejects atoms with $s_z = \frac{\hbar}{2}$. What is the intensity of the final beam with respect to that of the beam that survives the first measure? How do we orient the $\hat{\mathbf{n}}$ direction of the second device if the maximum final intensity is to be achieved?

Solution

After the transit in the first apparatus, the atoms are described by an eigenstate of S_z corresponding to the eigenvalue $+\hbar/2$. Using the results of exercise 4.2, this state can be written as a superimposition of eigenstates of $\mathbf{S} \cdot \hat{n}$ in the form

$$\left| S_z = +\frac{\hbar}{2} \right\rangle = c_+ \left| \mathbf{S} \cdot \hat{n} = +\frac{\hbar}{2} \right\rangle + c_- \left| \mathbf{S} \cdot \hat{n} = -\frac{\hbar}{2} \right\rangle,$$

where

$$c_+ = \left\langle \mathbf{S} \cdot \hat{n} = +\frac{\hbar}{2} \,\middle|\, S_z = +\frac{\hbar}{2} \right\rangle = \left(\cos \tfrac{\vartheta}{2} \;\; \sin \tfrac{\vartheta}{2} \right) \begin{pmatrix} 1 \\ 0 \end{pmatrix} = \cos \frac{\vartheta}{2},$$

$$c_- = \left\langle \mathbf{S} \cdot \hat{n} = -\frac{\hbar}{2} \,\middle|\, S_z = +\frac{\hbar}{2} \right\rangle = \left(\sin \tfrac{\vartheta}{2} \;\; -\cos \tfrac{\vartheta}{2} \right) \begin{pmatrix} 1 \\ 0 \end{pmatrix} = \sin \frac{\vartheta}{2}.$$

After the second measurement, the intensity of the beam will then be reduced by a factor of $\cos^2 \frac{\vartheta}{2}$, while each transmitted atom will be in the state

$$\left| \mathbf{S} \cdot \hat{n} = +\frac{\hbar}{2} \right\rangle = \begin{pmatrix} \cos \frac{\vartheta}{2} \\ \sin \frac{\vartheta}{2} \end{pmatrix} = \cos \frac{\vartheta}{2} \left| S_z = +\frac{\hbar}{2} \right\rangle + \sin \frac{\vartheta}{2} \left| S_z = -\frac{\hbar}{2} \right\rangle.$$

The third measurement further reduces the intensity of the beam by a factor of $\sin^2 \frac{\vartheta}{2}$, The third measure further reduces the intensity of the beam by a factor, so that the ratio between the intensity of the final beam and that of the beam that survives the first measure is given by

$$\cos^2 \frac{\vartheta}{2} \sin^2 \frac{\vartheta}{2} = \frac{1}{4} \sin^2 \vartheta.$$

This ratio is maximum for ϑ equal to $\pi/2$ or $3\pi/2$.

4.8 Energy Eigenstates of a System of Interacting Fermions

A system of three different particles of spin $\frac{1}{2}$ has the Hamiltonian

$$\mathcal{H} = K(\boldsymbol{\sigma}_1 \cdot \boldsymbol{\sigma}_2 + \boldsymbol{\sigma}_2 \cdot \boldsymbol{\sigma}_3 + \boldsymbol{\sigma}_3 \cdot \boldsymbol{\sigma}_1),$$

where K is a constant.

Determine the eigenvalues of $\mathcal{H}$, their degeneracy and the related eigenstates.

Solution

Having named **J** the system total spin and j its quantum number, we have

$$\mathcal{H} = K \frac{1}{2} \frac{4}{\hbar^2} [(\mathbf{S}_1 + \mathbf{S}_2 + \mathbf{S}_3)^2 - \mathbf{S}_1^2 - \mathbf{S}_2^2 - \mathbf{S}_3^2] = \frac{2K}{\hbar^2} \left[J^2 - \frac{9}{4} \hbar^2 \right].$$

Recall that, in the case of two spin $1/2$ particles, if we indicate the total spin eigenstate with $|j, j_z\rangle$ and the single particle spin eigenstates with $|\pm, \pm\rangle$, we have the following possible resulting states:

$$j_{12} = 0 \quad |0, 0\rangle = \frac{1}{\sqrt{2}}(|+, -\rangle - |-, +\rangle),$$

$$j_{12} = 1 \quad \begin{cases} |1, +1\rangle = |+, +\rangle \\ |1, 0\rangle = \frac{1}{\sqrt{2}}(|+, -\rangle + |-, +\rangle) \\ |1, -1\rangle = |-, -\rangle \end{cases}.$$

A third spin $1/2$ particle can be combined with the other pair, being either in the $j_{12} = 0$ state or in the $j_{12} = 1$ state. Combining a $j_{12} = 0$ pair with the third spin $1/2$ particle, we have two $j = 1/2$ states:

$$j_{123} = \frac{1}{2} \quad \begin{cases} |\frac{1}{2}, +\frac{1}{2}\rangle = \frac{1}{\sqrt{2}}(|+, -, +\rangle - |-, +, +\rangle) \\ |\frac{1}{2}, -\frac{1}{2}\rangle = \frac{1}{\sqrt{2}}(|+, -, -\rangle - |-, +, -\rangle) \end{cases}.$$

Using the Clebsh-Gordan coefficients (see, for example, http://pdg.lbl.gov/2002/clebrpp.pdf), we now combine the third spin $1/2$ particle with a pair in the state $j_{12} = 1$. This results in four $j = 3/2$ states:

$$j_{123} = \frac{3}{2} \quad \begin{cases} |\frac{3}{2}, +\frac{3}{2}\rangle = |m_{12} = +1, m_3 = +1/2\rangle = |+, +, +\rangle \\ |\frac{3}{2}, +\frac{1}{2}\rangle = \sqrt{\frac{1}{3}}|m_{12} = +1, m_3 = -1/2\rangle + \sqrt{\frac{2}{3}}|m_{12} = 0, m_3 = +1/2\rangle = \\ \quad = \sqrt{\frac{1}{3}}(|+, +, -\rangle + |+, -, +\rangle + |-, +, +\rangle) \\ |\frac{3}{2}, -\frac{1}{2}\rangle = \sqrt{\frac{1}{3}}|m_{12} = -1, m_3 = +1/2\rangle + \sqrt{\frac{2}{3}}|m_{12} = 0, m_3 = -1/2\rangle = \\ \quad = \sqrt{\frac{1}{3}}(|-, -, +\rangle + |+, -, -\rangle + |-, +, -\rangle) \\ |\frac{3}{2}, -\frac{3}{2}\rangle = |m_{12} = -1, m_3 = -1/2\rangle = |-, -, -\rangle \end{cases},$$

and two $j = 1/2$ states:

$$j_{123} = \frac{1}{2} \quad \begin{cases} |\frac{1}{2}, +\frac{1}{2}\rangle = \sqrt{\frac{2}{3}}|m_{12} = +1, m_3 = -1/2\rangle - \sqrt{\frac{1}{3}}|m_{12} = 0, m_3 = +1/2\rangle = \\ \quad = \sqrt{\frac{2}{3}}|+, +, -\rangle - \sqrt{\frac{1}{6}}|+, -, +\rangle - \sqrt{\frac{1}{6}}|-, +, +\rangle \\ |\frac{1}{2}, -\frac{1}{2}\rangle = \sqrt{\frac{1}{3}}|m_{12} = 0, m_3 = -1/2\rangle - \sqrt{\frac{2}{3}}|m_{12} = -1, m_3 = +1/2\rangle = \\ \quad = \sqrt{\frac{1}{6}}|+, -, -\rangle + \sqrt{\frac{1}{6}}|-, +, -\rangle - \sqrt{\frac{2}{3}}|-, -, +\rangle \end{cases}.$$

We note that we have determined four $j = 1/2$ states, instead of two. This may seem wrong, since the degeneration of $j = 1/2$ should be 2. However, it is easy to verify that the states obtained by the tensor product $j_{12} \otimes j_3 = 0 \otimes \frac{1}{2}$ are perpendicular to those obtained by $j_{12} \otimes j_3 = 1 \otimes \frac{1}{2}$. For example,

$$\frac{1}{\sqrt{2}}(\langle +, -, +| - \langle -, +, +|)(\sqrt{\frac{2}{3}} |+, +, -\rangle - \sqrt{\frac{1}{6}} |+, -, +\rangle - \sqrt{\frac{1}{6}} |-, +, +\rangle) =$$

$$= \frac{1}{2\sqrt{3}}(1 - 1) = 0.$$

After all, the independent states must be 8, since we started from a vector space of dimension 8 ($j_1 \otimes j_2 \otimes j_3 = \frac{1}{2} \otimes \frac{1}{2} \otimes \frac{1}{2}$).

Ultimately, the Hamiltonian eigenvalues are:

$$j_{123} = \frac{1}{2} \qquad E_0 = \frac{2K}{\hbar^2} \left[\frac{3}{4}\hbar^2 - \frac{9}{4}\hbar^2 \right] = -3K \text{ with degeneracy 4,}$$

$$j_{123} = \frac{3}{2} \qquad E_1 = \frac{2K}{\hbar^2} \left[\frac{15}{4}\hbar^2 - \frac{9}{4}\hbar^2 \right] = +3K \text{ with degeneracy 4.}$$

4.9   Spin Measurements on a Fermion

Consider a spin particle $\frac{1}{2}$ and suppose you are measuring the sum of x and z spin components $S_x + S_z$. What are the possible results of the measurement? If you subsequently measure S_y, what is the probability of finding the value $+\frac{\hbar}{2}$?

Solution

We notice that

$$S_x + S_z = \sqrt{2}\, \mathbf{S} \cdot \mathbf{n},$$

where $\mathbf{n}$ is the unit vector in the direction of the bisector of the xz-plan ($\vartheta = \pi/4$, $\varphi = 0$).

Because of space isotropy, it is obvious that the eigenvalues of $\mathbf{S} \cdot \mathbf{n}$ are $\pm\hbar/2$ for any $\mathbf{n}$. Therefore, the possible results of the measurement are $\pm\hbar/\sqrt{2}$.

After the measurement, the particle spin will be in the xz plane, so the probability of finding either of the two possible eigenvalues of S_y will be $1/2$. In fact, as we have seen in the general case in Problem 3.4, in an eigenstate of a component of the angular momentum, the components perpendicular to it have a null expectation value. Since the eigenvalues of S_y, as of any spin component, are opposite, it follows that the probabilities relative to the two measures are equal.

Chapter 5
Time Evolution

5.1 Two-Level System (I)

The Hamiltonian of a two-level quantum system can be written as

$$\mathcal{H} = -\frac{1}{2}\hbar\omega(|0\rangle\langle 0| - |1\rangle\langle 1|),$$

where $|0\rangle$ and $|1\rangle$ are its orthonormal eigenkets corresponding to the eigenvalues $-\hbar\omega/2$ and $+\hbar\omega/2$, respectively. Consider the linear operator $a = |0\rangle\langle 1|$ and its hermitian conjugate $a^\dagger$.

(a) Prove that the following relationships are valid:

$$\{a, a^\dagger\} = aa^\dagger + a^\dagger a = 1, \quad a^2 = a^{\dagger 2} = 0,$$

$$[\mathcal{H}, a] = -\hbar\omega a, \quad [\mathcal{H}, a^\dagger] = +\hbar\omega a^\dagger,$$

that the operator $N = aa^\dagger$ has eigenvalues 0 and 1 and that its eigenkets are the basis kets. Express the Hamiltonian in terms of N and the identity I.

(b) Assume that, at instant $t = 0$, the system is in the eigenstate of the hermitian operator $A = a + a^\dagger$ corresponding to the eigenvalue 1 and determine the expectation values of A and A^2 and the uncertainty $\langle(\Delta A)^2\rangle$ as a function of time t.

(c) Consider the other hermitian operator $B = -i(a - a^\dagger)$; also, determine $\langle B\rangle$, $\langle B^2\rangle$ and $\langle(\Delta B)^2\rangle$ as a function of time. Verify the uncertainty principle between A and B.

Solution

(a) From the eigenvalue equation for $\mathcal{H}$, we have

$$\mathcal{H}|0\rangle = -\frac{\hbar\omega}{2}|0\rangle, \quad \mathcal{H}|1\rangle = +\frac{\hbar\omega}{2}|1\rangle,$$

© Springer Nature Switzerland AG 2019
L. Angelini, *Solved Problems in Quantum Mechanics*, UNITEXT for Physics,
https://doi.org/10.1007/978-3-030-18404-9_5

As states $|0\rangle$ and $|1\rangle$ are an orthonormal set,

$$a = |0\rangle\langle 1|, \quad a^\dagger = |1\rangle\langle 0|,$$
$$\{a, a^\dagger\} = aa^\dagger + a^\dagger a = |0\rangle\langle 1|1\rangle\langle 0| + |1\rangle\langle 0|0\rangle\langle 1| = |0\rangle\langle 0| + |1\rangle\langle 1| = 1,$$
$$a^2 = |0\rangle\langle 1|0\rangle\langle 1| = 0,$$
$$a^{\dagger 2} = |1\rangle\langle 0|1\rangle\langle 0| = 0,$$

$$[\mathcal{H}, a] = -\frac{1}{2}\hbar\omega \left(|0\rangle\langle 0|0\rangle\langle 1| - |1\rangle\langle 1|0\rangle\langle 1| - |0\rangle\langle 1|0\rangle\langle 0| + |0\rangle\langle 1|1\rangle\langle 1|\right) =$$
$$= -\hbar\omega|0\rangle\langle 1| = -\hbar\omega a,$$

$$[\mathcal{H}, a^\dagger] = -\frac{1}{2}\hbar\omega \left(|0\rangle\langle 0|1\rangle\langle 0| - |1\rangle\langle 1|1\rangle\langle 0| - |1\rangle\langle 0|0\rangle\langle 0| + |1\rangle\langle 0|1\rangle\langle 1|\right) =$$
$$= \hbar\omega|1\rangle\langle 0| = \hbar\omega a^\dagger,$$

$$N = aa^\dagger = |0\rangle\langle 1|1\rangle\langle 0| = |0\rangle\langle 0|$$
$$N|0\rangle = |0\rangle\langle 0|0\rangle = |0\rangle \quad \Rightarrow \quad |0\rangle \text{ eigenket with eigenvalue 1}$$
$$N|1\rangle = |0\rangle\langle 0|1\rangle = 0 = 0|1\rangle \quad \Rightarrow \quad |1\rangle \text{ eigenket with eigenvalue 0,}$$
$$\mathcal{H} = -\frac{1}{2}\hbar\omega(|0\rangle\langle 0| - |1\rangle\langle 1| + |0\rangle\langle 0| - |0\rangle\langle 0|) = -\frac{1}{2}\hbar\omega(2N - I) =$$
$$= \hbar\omega(\frac{I}{2} - N).$$

(b) Consider the state vector

$$|\psi\rangle = \frac{1}{\sqrt{2}} (|0\rangle + |1\rangle) .$$

$|\psi\rangle$ is the eigenket of A corresponding to eigenvalue 1. In effect,

$$A|\psi\rangle = (|0\rangle\langle 1| + |1\rangle\langle 0|)\frac{1}{\sqrt{2}} (|0\rangle + |1\rangle) = \frac{1}{\sqrt{2}} (0 + |0\rangle + |1\rangle + 0) = 1 \cdot |\psi\rangle.$$

At time t the state vector $|\psi\rangle$ will be given by

$$|\psi(t)\rangle = \frac{1}{\sqrt{2}} \left(e^{i\frac{\omega}{2}t}|0\rangle + e^{-i\frac{\omega}{2}t}|1\rangle\right) .$$

We note that

$$[\mathcal{H}, A] = [\mathcal{H}, a] + [\mathcal{H}, a^\dagger] = -\hbar\omega(a - a^\dagger) = -i\hbar\omega B \neq 0,$$
$$[\mathcal{H}, B] = -i \left([\mathcal{H}, a] - [\mathcal{H}, a^\dagger]\right) = i\hbar\omega(a + a^\dagger) = i\hbar\omega A \neq 0,$$

($B = -i(a - a^\dagger)$ is defined in the third question). A and B do not commute with $\mathcal{H}$, thus their expectation values depend on time. Indeed, we find that

$$\langle A \rangle_\psi = \frac{1}{2} \left(\langle 0 | e^{-i\frac{\omega}{2}t} + \langle 1 | e^{i\frac{\omega}{2}t} \right) (a + a^\dagger) \left(e^{i\frac{\omega}{2}t} | 0 \rangle + e^{-i\frac{\omega}{2}t} | 1 \rangle \right) =$$

$$= \frac{1}{2} \left(\langle 0 | e^{-i\frac{\omega}{2}t} + \langle 1 | e^{i\frac{\omega}{2}t} \right) \left(e^{i\frac{\omega}{2}t} | 1 \rangle + e^{-i\frac{\omega}{2}t} | 0 \rangle \right) =$$

$$= \frac{1}{2} \left(e^{i\omega t} + e^{-i\omega t} \right) = \cos \omega t,$$

$$\langle A^2 \rangle_\psi = \langle (a + a^\dagger)(a + a^\dagger) \rangle_\psi = \langle aa^\dagger + a^\dagger a \rangle_\psi =$$

$$= \langle \{a, a^\dagger\} \rangle_\psi = \langle I \rangle_\psi = 1,$$

$$\langle (\Delta A)^2 \rangle_\psi = \langle A^2 \rangle_\psi - \langle A \rangle_\psi^2 = 1 - \cos^2 \omega t = \sin^2 \omega t.$$

(c) Similarly, the following relationships are obtained for B:

$$\langle B \rangle_\psi = -i\frac{1}{2} \left(\langle 0 | e^{-i\frac{\omega}{2}t} + \langle 1 | e^{i\frac{\omega}{2}t} \right) (a - a^\dagger) \left(e^{i\frac{\omega}{2}t} | 0 \rangle + e^{-i\frac{\omega}{2}t} | 1 \rangle \right) =$$

$$= -i\frac{1}{2} \left(\langle 0 | e^{-i\frac{\omega}{2}t} + \langle 1 | e^{i\frac{\omega}{2}t} \right) \left(e^{-i\frac{\omega}{2}t} | 0 \rangle - e^{+i\frac{\omega}{2}t} | 1 \rangle \right) =$$

$$= -i\frac{1}{2} \left(e^{-i\omega t} - e^{i\omega t} \right) = -\sin \omega t,$$

$$\langle B^2 \rangle_\psi = \langle i^2 (a - a^\dagger)(a - a^\dagger) \rangle_\psi = -\langle -aa^\dagger - a^\dagger a \rangle_\psi =$$

$$= \langle \{a, a^\dagger\} \rangle_\psi = \langle I \rangle_\psi = 1,$$

$$\langle (\Delta B)^2 \rangle_\psi = \langle B^2 \rangle_\psi - \langle B \rangle_\psi^2 = 1 - \sin^2 \omega t = \cos^2 \omega t.$$

As for the uncertainty relationship between A and B, we have:

$$\Delta A \cdot \Delta B = |\sin \omega t| \cdot |\cos \omega t|.$$

Recall that, whatever the state, the following relationship must hold:

$$\Delta A \cdot \Delta B \geq \frac{1}{2} |\langle [A, B] \rangle|.$$

Let us check its validity in this case as well:

$$[A, B] = -i[a + a^\dagger, a - a^\dagger] =$$
$$= -i \left([a, a] + [a^\dagger, a] - [a, a^\dagger] - [a^\dagger, a^\dagger] \right) =$$
$$= -i \left(a^\dagger a - aa^\dagger - aa^\dagger + a^\dagger a \right) =$$
$$= 2i \left(aa^\dagger - a^\dagger a \right) = 2i \left(|0\rangle\langle 1|1\rangle\langle 0| - |1\rangle\langle 0|0\rangle\langle 1| \right) =$$
$$= 2i \left(|0\rangle\langle 0| - |1\rangle\langle 1| \right) = -\frac{4i}{\hbar\omega} \mathcal{H},$$

$$\langle [A, B] \rangle_\psi = -\frac{4i}{\hbar\omega}\frac{1}{2} \left(\langle 0|e^{-i\frac{\omega}{2}t} + \langle 1|e^{i\frac{\omega}{2}t} \right) \mathcal{H} \left(e^{i\frac{\omega}{2}t}|0\rangle + e^{-i\frac{\omega}{2}t}|1\rangle \right) =$$

$$= -i \left(\langle 0|e^{-i\frac{\omega}{2}t} + \langle 1|e^{i\frac{\omega}{2}t} \right) \left(-e^{i\frac{\omega}{2}t}|0\rangle + e^{-i\frac{\omega}{2}t}|1\rangle \right) =$$

$$= -i(-1 + 1) = 0.$$

The uncertainty principle is verified, in that

$$|\sin \omega t| \cdot |\cos \omega t| \geq 0.$$

5.2   Two-Level System (II)

Consider a two-state system whose Hamiltonian is

$$\mathcal{H} = E_0|1\rangle\langle 1| + \sqrt{2}E_0|1\rangle\langle 2| + \sqrt{2}E_0|2\rangle\langle 1|,$$

where $\{|1\rangle, |2\rangle\}$ is an orthonormal basis in the system's Hilbert space.
If the system is initially in state $|1\rangle$, how likely is it that it will be in state $|2\rangle$ at time
t? Determine the period of oscillations between states $|1\rangle$ and $|2\rangle$.

Solution

In the $\{|1\rangle, |2\rangle\}$ representation, the Hamiltonian becomes a matrix:

$$\mathcal{H} = \begin{pmatrix} E_0 & \sqrt{2}E_0 \\ \sqrt{2}E_0 & 0 \end{pmatrix} = E_0\mathcal{H}' \quad \text{where} \quad \mathcal{H}' = \begin{pmatrix} 1 & \sqrt{2} \\ \sqrt{2} & 0 \end{pmatrix}.$$

From the secular equation, we obtain the eigenvalues of $\mathcal{H}'$:

$$\det(\mathcal{H}' - \lambda I) = \lambda^2 - \lambda - 2 = 0 \quad \Rightarrow \quad \lambda = -1, 2,$$

and the corresponding $\mathcal{H}$'s eigenvalues $E_1 = -E_0$ and $E_2 = 2E_0$. Eigenket $|E_1\rangle$ is
obtained by:

$$\begin{pmatrix} 1 & \sqrt{2} \\ \sqrt{2} & 0 \end{pmatrix} \begin{pmatrix} a \\ b \end{pmatrix} = -1 \begin{pmatrix} a \\ b \end{pmatrix},$$

from which we get, normalizing,

$$|E_1\rangle = \frac{1}{\sqrt{3}} \begin{pmatrix} 1 \\ -\sqrt{2} \end{pmatrix} = \frac{1}{\sqrt{3}}|1\rangle - \sqrt{\frac{2}{3}}|2\rangle.$$

Similarly, we obtain the eigenket $|E_2\rangle$:

$$|E_2\rangle = \frac{1}{\sqrt{3}}\begin{pmatrix}\sqrt{2}\\1\end{pmatrix} = \sqrt{\frac{2}{3}}|1\rangle + \frac{1}{\sqrt{3}}|2\rangle.$$

By reversing these relationships, we have

$$|1\rangle = \frac{1}{\sqrt{3}}|E_1\rangle + \sqrt{\frac{2}{3}}|E_2\rangle,$$

$$|2\rangle = -\sqrt{\frac{2}{3}}|E_1\rangle + \frac{1}{\sqrt{3}}|E_2\rangle.$$

Initially, the system state is

$$|\psi(t=0)\rangle = |1\rangle = \frac{1}{\sqrt{3}}|E_1\rangle + \sqrt{\frac{2}{3}}|E_2\rangle;$$

at instant t, it will be evolved in

$$|\psi(t)\rangle = \frac{1}{\sqrt{3}}e^{i\frac{E_0 t}{\hbar}}|E_1\rangle + \sqrt{\frac{2}{3}}e^{-i\frac{2E_0 t}{\hbar}}|E_2\rangle =$$

$$= \frac{1}{3}e^{i\frac{E_0 t}{\hbar}}\left[\left(1 + 2e^{-i\frac{3E_0 t}{\hbar}}\right)|1\rangle - \sqrt{2}\left(1 - e^{-i\frac{3E_0 t}{\hbar}}\right)|2\rangle\right].$$

The probabilities of finding the system in one of the basis vectors at the time t are

$$P_{|1\rangle}(t) = \frac{1}{9}\left|1 + 2e^{-i\frac{3E_0 t}{\hbar}}\right|^2 = \frac{1}{9}\left(5 + 4\cos\frac{3E_0 t}{\hbar}\right),$$

$$P_{|2\rangle}(t) = \frac{2}{9}\left|1 - e^{-i\frac{3E_0 t}{\hbar}}\right|^2 = \frac{4}{9}\left(1 - \cos\frac{3E_0 t}{\hbar}\right).$$

These probabilities oscillate with frequency $\omega = 3E_0/\hbar$. Therefore, the desired period is

$$T = \frac{2\pi}{\omega} = \frac{2\pi\hbar}{3E_0} = \frac{h}{3E_0}.$$

5.3 Two-Level System (III)

Consider a two-state system and the basis composed by $|\psi_1\rangle$ and $|\psi_2\rangle$, which are eigenkets of the Hamiltonian H_0, corresponding to eigenvalues E_1 and E_2, respectively,

$$\mathcal{H}_0|\psi_1\rangle = E_1|\psi_1\rangle,$$
$$\mathcal{H}_0|\psi_2\rangle = E_2|\psi_2\rangle,$$
$$\langle\psi_i|\psi_j\rangle = \delta_{i,j} \quad i, j = 1, 2.$$

Consider a new system with Hamiltonian $\mathcal{H}_0 + W$, where the coupling term W, in the basis $\{|\psi_1\rangle, |\psi_2\rangle\}$, is given by the 2×2 matrix W_{ij} with $W_{11} = W_{22} = 0$ and $W_{12} = W_{21} = w$, where w is a positive real constant.

(a) Determine how the Hamiltonian's eigenstates and eigenvalues change as a result of this coupling.

(b) If, at instant $t = 0$, the second system is known with certainty to be in state $|\psi_1\rangle$, at which instants (if they exist) will the system be in the same condition again? Calculate the probability of finding the system in state $|\psi_2\rangle$ at time t.

Solution

(a) In the $\{|\psi_1\rangle, |\psi_2\rangle\}$ basis representation, we have

$$\mathcal{H}_0 = \begin{pmatrix} E_1 & 0 \\ 0 & E_2 \end{pmatrix}, \quad W = \begin{pmatrix} 0 & w \\ w & 0 \end{pmatrix}, \quad \mathcal{H} = \mathcal{H}_0 + W = \begin{pmatrix} E_1 & w \\ w & E_2 \end{pmatrix},$$

where, in order not to weigh down the notation, we have used the same symbols of operators and kets for the matrices corresponding to them. $\mathcal{H}$'s eigenvalues are obtained by $(E_1 - \lambda)(E_2 - \lambda) - W^2 = 0$:

$$\lambda_\pm = \frac{E_1 + E_2 \pm \sqrt{(E_1 - E_2)^2 + 4w^2}}{2}.$$

Eigenvectors $|\lambda_\pm\rangle$ are obtained from equations

$$\mathcal{H}|\lambda_\pm\rangle = \lambda_\pm|\lambda_\pm\rangle.$$

It is easily found, appropriately choosing the phases and imposing normalization, that

$$|\lambda_+\rangle \equiv \begin{pmatrix} \langle\psi_1|\lambda_+\rangle \\ \langle\psi_2|\lambda_+\rangle \end{pmatrix} = \frac{1}{\sqrt{(\lambda_+ - E_1)^2 + w^2}} \begin{pmatrix} w \\ \lambda_+ - E_1 \end{pmatrix},$$

$$|\lambda_-\rangle \equiv \begin{pmatrix} \langle\psi_1|\lambda_-\rangle \\ \langle\psi_2|\lambda_-\rangle \end{pmatrix} = \frac{1}{\sqrt{(\lambda_- - E_1)^2 + w^2}} \begin{pmatrix} w \\ \lambda_- - E_1 \end{pmatrix}.$$

(b) The system will certainly be in state $|\psi_1\rangle$ when the probability of finding it in state $|\psi_2\rangle$ will be null. The state at time $t > 0$ is

$$|\psi(t)\rangle = e^{-i\frac{\mathcal{H}t}{\hbar}} |\psi_1\rangle =$$

$$= e^{-i\frac{\lambda_+ t}{\hbar}} |\lambda_+\rangle\langle\lambda_+|\psi_1\rangle + e^{-i\frac{\lambda_- t}{\hbar}} |\lambda_-\rangle\langle\lambda_-|\psi_1\rangle =$$

$$= e^{-i\frac{\lambda_+ t}{\hbar}} \frac{w}{(\lambda_+ - E_1)^2 + w^2} \begin{pmatrix} w \\ \lambda_+ - E_1 \end{pmatrix} +$$

$$+ e^{-i\frac{\lambda_- t}{\hbar}} \frac{w}{(\lambda_- - E_1)^2 + w^2} \begin{pmatrix} w \\ \lambda_- - E_1 \end{pmatrix}.$$

Introducing the quantities

$$\Delta = \frac{E_2 - E_1}{2}, \qquad \Sigma = \frac{E_2 + E_1}{2} \qquad \text{and} \qquad \alpha = \sqrt{\Delta^2 + w^2},$$

we obtain

$$\lambda_\pm - E_1 = \Delta \pm \alpha \quad \text{and} \quad \lambda_\pm = \Sigma \pm \alpha.$$

The probability of finding the system in state $|\psi_2\rangle$ at time t is given by the square modulus of $\langle\psi_2|\psi\rangle$, which, in terms of the new variables, is equal to

$$\langle\psi_2|\psi\rangle = w e^{-i\frac{\Sigma t}{\hbar}} \left(c_1 e^{-i\frac{\alpha t}{\hbar}} + c_2 e^{i\frac{\alpha t}{\hbar}} \right) = w e^{-i\frac{\Sigma t}{\hbar}} \left[(c_1 + c_2) \cos\frac{\alpha t}{\hbar} - i(c_1 - c_2) \sin\frac{\alpha t}{\hbar} \right],$$

where

$$c_1 = \frac{\Delta + \alpha}{(\Delta + \alpha)^2 + w^2}, \qquad c_2 = \frac{\Delta - \alpha}{(\Delta - \alpha)^2 + w^2}.$$

After short calculations, we find that

$$c_1 + c_2 = 0 \qquad \text{and} \qquad c_1 - c_2 = \frac{1}{\alpha},$$

and, consequently,

$$\langle\psi_2|\psi\rangle = -i \frac{w}{\alpha} e^{-i\frac{\Sigma t}{\hbar}} \sin\frac{\alpha t}{\hbar}.$$

Ultimately, the answer to the problem's question is

$$P\big(|\psi(t)\rangle = |\psi_2\rangle\big) = |\langle\psi_2|\psi\rangle|^2 = \frac{w^2}{\Delta^2 + w^2} \sin^2\frac{\alpha}{\hbar} t.$$

This probability is null and $|\psi(t)\rangle = |\psi_1\rangle$ when

$$t = \frac{n\pi\hbar}{\alpha} \qquad \text{con} \quad n = 0, 1, 2, \dots$$

5.4   Two-Level System (IV)

The Hamiltonian of a two-level quantum system is described by the following operator (in the appropriate units of measurement):

$$\mathcal{H}|1\rangle = |1\rangle + \frac{1+i}{\sqrt{2}}|2\rangle, \qquad \mathcal{H}|2\rangle = \frac{1-i}{\sqrt{2}}|1\rangle + |2\rangle,$$

where $|1\rangle$ and $|2\rangle$ are the normalized eigenkets of another hermitian operator $\mathcal{A}$:

$$\mathcal{A}|1\rangle = \sqrt{2}|1\rangle, \qquad \mathcal{A}|2\rangle = -\sqrt{2}|2\rangle.$$

At the instant $t = 0$, a measure of the observable associated with the operator $\mathcal{A}$ is performed and the value $-\sqrt{2}$ is found.

(a) Immediately afterwards, an energy measurement is performed; what is the probability of finding the system in its ground state?
(b) How does this probability change by performing the measurement after a finite interval T?
(c) If no energy measurement is taken, at which time will the system be in state $|2\rangle$?

Solution

In the $\mathcal{A}$'s eigenkets representation, to $\mathcal{H}$ there corresponds a matrix, which we call by the same name, having elements $\mathcal{H}_{j,k} = \langle j|C|k\rangle$,

$$\mathcal{H} = \begin{pmatrix} 1 & \frac{1-i}{\sqrt{2}} \\ \frac{1+i}{\sqrt{2}} & 1 \end{pmatrix}.$$

Its eigenvalues are $E_\pm = 2, 0$. The corresponding eigenvectors are, respectively,

$$\begin{pmatrix} \frac{1}{\sqrt{2}} \\ \frac{1+i}{2} \end{pmatrix} \quad \text{and} \quad \begin{pmatrix} \frac{1}{\sqrt{2}} \\ -\frac{1+i}{2} \end{pmatrix}.$$

Therefore, we can write

$$|E_-\rangle = \frac{1}{\sqrt{2}}|1\rangle - \frac{1+i}{2}|2\rangle,$$

$$|E_+\rangle = \frac{1}{\sqrt{2}}|1\rangle + \frac{1+i}{2}|2\rangle.$$

By reversing these expressions, we find that

$$|1\rangle = \frac{1}{\sqrt{2}} (|E_+\rangle + |E_-\rangle),$$

$$|2\rangle = \frac{1-i}{2} (|E_+\rangle - |E_-\rangle).$$

(a) At time $t = 0$, the system's state is

$$|\psi(0)\rangle = |2\rangle = \frac{1-i}{2} (|E_+\rangle - |E_-\rangle),$$

and therefore

$$P(E = E_- = 0, t = 0) = \left| -\frac{1-i}{2} \right|^2 = \frac{1}{2}.$$

(b) At time $t = T$, the state has become

$$|\psi(T)\rangle = e^{-i\frac{\mathcal{H}T}{\hbar}} |\psi(0)\rangle = \frac{1-i}{2} \left(e^{-i\frac{2T}{\hbar}} |E_+\rangle - e^{-i\frac{0T}{\hbar}} |E_-\rangle \right) =$$

$$= \frac{1-i}{2} \left(e^{-i\frac{2T}{\hbar}} |E_+\rangle - |E_-\rangle \right),$$

implying that

$$P(E = E_- = 0, t = T) = \left| \frac{1-i}{2} \right|^2 = \frac{1}{2}$$

independent of T.

(c) In order for the system to be certainly in state $|2\rangle$ at time t, the following must occur:

$$\frac{1-i}{2} \left(e^{-i\frac{2t}{\hbar}} |E_+\rangle - |E_-\rangle \right) = \frac{1-i}{2} (|E_+\rangle - |E_-\rangle),$$

from which we get

$$\frac{2t}{\hbar} = 2n\pi \quad \Rightarrow \quad t = n\pi\hbar \quad \text{with } n = 1, 2, \dots$$

Note that t has the same dimension of $\hbar$ due to the fact that $\mathcal{H}$ is dimensionless.

5.5 Time-Evolution of a Free Particle

Consider a free particle in one dimension.

(a) Calculate the expression of the evolution operator (propagator) in the coordinate representation.

(b) Suppose that, at instant $t' = 0$, the particle is described by a Gaussian wave packet

$$\psi(x, t' = 0) = \frac{1}{\sqrt[4]{2\Delta^2 \pi}} e^{\iota k_0 x - \frac{x^2}{4\Delta^2}} \tag{5.1}$$

corresponding to a position uncertainty $\Delta x = \Delta$ and a momentum uncertainty $\Delta p = \frac{\hbar}{2\Delta}$. Calculate the wave function at a later time t and show that the position uncertainty grows over time (wave packet enlargement).

Solution

(a) Remember that the propagator, when the Hamiltonian does not depend on time, is given by

$$U(t, t') = e^{-\iota \frac{\mathcal{H}(t-t')}{\hbar}}.$$

In the position representation, for a free particle in one dimension, it assumes the expression

$$U(x, t, x', t') = \langle x | e^{-\iota \frac{\mathcal{H}(t-t')}{\hbar}} | x' \rangle = \int_{-\infty}^{+\infty} dp \langle x | e^{-\iota \frac{\mathcal{H}(t-t')}{\hbar}} | p \rangle \langle p | x' \rangle =$$

$$= \frac{1}{2\pi\hbar} \int_{-\infty}^{+\infty} dp \; e^{\iota \frac{p(x-x')}{\hbar} - \iota \frac{p^2(t-t')}{2m\hbar}}.$$

Using the well-known result for the Gaussian integral (A.4), we easily obtain

$$U(x, t, x', t') = \sqrt{\frac{m}{2\pi \hbar \iota (t - t')}} e^{\iota \frac{m(x-x')^2}{2\hbar(t-t')}}.$$

(b) We apply the previous result to the Gaussian wave packet (5.1).
    At time t, the wave function will be

$$\psi(x, t) = \int_{-\infty}^{+\infty} dx' \; U(x, t, x', 0) \psi(x', 0) =$$

$$= \frac{1}{\sqrt[4]{2\Delta^2\pi}} \sqrt{\frac{m}{2\pi \hbar \iota t}} \int_{-\infty}^{+\infty} dx' \; e^{\iota k_0 x' - \frac{x'^2}{4\Delta^2} + \iota \frac{m(x-x')^2}{2\hbar t}}.$$

This integral can be calculated using formula (A.4) again, placing

$$\alpha = \frac{1}{4\Delta^2} - \frac{\iota m}{2\hbar t} \quad \text{and} \quad \beta = \iota \left(k_0 - \frac{mx}{\hbar t} \right)$$

and obtaining

$$\psi(x,t) = \frac{1}{\sqrt[4]{2\Delta^2\pi}} \sqrt{\frac{m}{2\pi\hbar\iota t}} \sqrt{\frac{\pi}{\alpha}} \; e^{\frac{\iota m x^2}{2\hbar t}} \; e^{\frac{\beta^2}{4\alpha}} =$$

$$= \frac{1}{\sqrt{\sqrt{2\Delta^2\pi}\left(1+\iota\frac{\hbar t}{2m\Delta^2}\right)}} \; e^{\frac{\beta^2}{4\alpha}+\frac{\iota m x^2}{2\hbar t}}.$$

We explicate the exponent:

$$\frac{\beta^2}{4\alpha} + \frac{\iota m x^2}{2\hbar t} = \frac{-\left(\frac{mx}{\hbar t}-k_0\right)^2}{4\Delta^2\left(1+\iota\frac{\hbar t}{2m\Delta^2}\right)\frac{m}{2\iota\hbar t\Delta^2}} + \frac{\iota m x^2}{2\hbar t} =$$

$$= \frac{-\left(\frac{mx}{\hbar t}-k_0\right)^2\frac{2\iota\hbar t\Delta^2}{m} + \frac{\iota m x^2}{2\hbar t}4\Delta^2\left(1+\iota\frac{\hbar t}{2m\Delta^2}\right)}{4\Delta^2\left(1+\iota\frac{\hbar t}{2m\Delta^2}\right)} =$$

$$= \frac{-x^2 - \frac{2\iota\hbar t\Delta^2}{m}k_0\left(k_0-\frac{2mx}{\hbar t}\right)}{4\Delta^2\left(1+\iota\frac{\hbar t}{2m\Delta^2}\right)} =$$

$$= \frac{-x^2 + \frac{4\iota\Delta^2}{\hbar}p_0\left(x-\frac{p_0 t}{2m}\right)}{4\Delta^2\left(1+\iota\frac{\hbar t}{2m\Delta^2}\right)} =$$

$$= \frac{-x^2 + 2\frac{p_0 t}{m}x - \left(\frac{p_0 t}{m}\right)^2}{4\Delta^2\left(1+\iota\frac{\hbar t}{2m\Delta^2}\right)} + \frac{-2\frac{p_0 t}{m}x + \left(\frac{p_0 t}{m}\right)^2 + \frac{4\iota\Delta^2}{\hbar}p_0\left(x-\frac{p_0 t}{2m}\right)}{4\Delta^2\left(1+\iota\frac{\hbar t}{2m\Delta^2}\right)} =$$

$$= \frac{(x-p_0 t/m)^2}{4\Delta^2\left(1+\iota\frac{\hbar t}{2m\Delta^2}\right)} + \iota\frac{p_0}{\hbar}\left(x-\frac{p_0 t}{2m}\right).$$

The final expression for the Gaussian packet at time t is

$$\psi(x,t) = \frac{1}{\sqrt{\sqrt{2\Delta^2\pi}\left(1+\iota\frac{\hbar t}{2m\Delta^2}\right)}} \; e^{-\frac{(x-p_0 t/m)^2}{4\Delta^2\left(1+\iota\frac{\hbar t}{2m\Delta^2}\right)}+\iota\frac{p_0}{\hbar}\left(x-\frac{p_0 t}{2m}\right)}.$$

This expression shows that the wave packet moves like a plane wave of energy $E = p_0^2/2m$, modulated by a Gaussian packet whose peak value, which is also the position expectation value, moves according to $x = p_0 t/m$. The momentum expectation value remains $\langle p \rangle = p_0 = \hbar k_0$. Finally, the position probability distribution, which is obtained from the wave function's square modulus, shows that the position uncertainty changes over time with the law

$$\Delta x(t) = \Delta\sqrt{1+\frac{\hbar^2 t^2}{4m\Delta^4}},$$

giving rise to the phenomenon known as wave packet enlargement.

5.6 Particle Confined on a Segment (I)

At the instant $t = 0$, a particle, constrained to move along a segment of length L, is in a state in which an energy measure can supply, with equal probability, only two values: the lower E_1 and the next higher $E_2 = 4E_1$.

(a) Write the expression of the normalized wave function (containing an arbitrary parameter).
(b) Determine this parameter, knowing that, at the instant $t = 0$, the particle momentum expectation value is $\langle p \rangle = \frac{4}{3} \frac{\hbar}{L}$.
(c) Determine the next time instant at which the momentum expectation value vanishes.

Solution

For this system, energy eigenvalues and corresponding eigenfunctions are given by:

$$E_n = \frac{\hbar^2 \pi^2 n^2}{2mL^2}, \quad \psi_n(x) = \sqrt{\frac{2}{L}} \sin \frac{n\pi x}{L}, \quad n = 1, 2, 3, \ldots.$$

(a) At $t = 0$, the particle state is a superposition of $\psi_1(x)$ and $\psi_2(x)$:

$$\psi(x) = c_1 \psi_1(x) + c_2 \psi_2(x).$$

Since the probabilities of finding E_1 and E_2 are the same, we have

$$|c_1|^2 = |c_2|^2.$$

Neglecting an arbitrary overall phase, we can write

$$\psi(x) = \frac{1}{\sqrt{2}} \left(\psi_1(x) + e^{i\alpha} \psi_2(x) \right).$$

(b) Let us calculate the momentum expectation value. As we have seen in Problems (1.4) and (2.4), this is null for a particle in an eigenstate of the discrete spectrum. Therefore, we have

$$\langle p \rangle_{t=0} = \frac{1}{2} \left[\langle \psi_1 | p | \psi_1 \rangle + \langle \psi_2 | p | \psi_2 \rangle + e^{i\alpha} \langle \psi_1 | p | \psi_2 \rangle + e^{-i\alpha} \langle \psi_2 | p | \psi_1 \rangle \right] =$$
$$= \frac{1}{2} \left[e^{i\alpha} I_1 + e^{-i\alpha} I_2 \right],$$

where $I_2^* = I_1$. The latter is given by

$$I_1 = \frac{2}{L}\frac{\hbar}{i}\int_0^L dx \left(\sin\frac{\pi x}{L}\right)\frac{2\pi}{L}\cos\frac{2\pi x}{L} = \frac{4\hbar}{iL}\int_0^\pi dt\,\sin t\,\cos 2t =$$

$$= \frac{2\hbar}{iL}\int_{-1}^1 dz\,(2z^2 - 1) = -\frac{8\hbar}{3iL}.$$

Finally,

$$\langle p\rangle_{t=0} = \frac{1}{2}\frac{8\hbar}{3iL}\left[-e^{i\alpha} + e^{-i\alpha}\right] = -\frac{8\hbar}{3L}\sin\alpha.$$

By imposing the required condition, two different determinations of α are obtained:

$$\sin\alpha = -\frac{1}{2} \quad \Rightarrow \quad \alpha = \alpha_1 = -\frac{\pi}{6} \quad \text{or} \quad \alpha = \alpha_2 = \pi + \frac{\pi}{6}.$$

(c) At a later time t, we will have

$$\psi(x, t) = \frac{1}{\sqrt{2}}\left(e^{-i\frac{E_1 t}{\hbar}}\psi_1(x) + e^{i\alpha}e^{-i\frac{E_2 t}{\hbar}}\psi_2(x)\right)$$

and the momentum expectation value will be given by

$$\langle p\rangle_t = \frac{1}{2}\left[e^{i\alpha}e^{-i\frac{E_2-E_1}{\hbar}t}\langle\psi_1|p|\psi_2\rangle + e^{-i\alpha}e^{i\frac{E_2-E_1}{\hbar}t}\langle\psi_2|p|\psi_1\rangle\right] =$$

$$= \frac{1}{2}\frac{8\hbar}{3iL}\left[-e^{i(\alpha-\omega t)} + e^{-i(\alpha-\omega t)}\right] = -\frac{8\hbar}{3L}\sin(\alpha + \omega t),$$

where we defined

$$\omega = \frac{E_2 - E_1}{\hbar}.$$

Depending on the determination of α, the momentum expectation value will take zero for the first time when $\omega t = \alpha_1 + \pi$ or $\omega t = \alpha_2 - \pi$, that is, at time given by, respectively for the two cases,

$$t = \frac{5\pi}{6\omega} \quad \text{or} \quad t = \frac{\pi}{6\omega}.$$

5.7 Particle Confined on a Segment (II)

A particle of mass m is confined on the segment $0 \le x \le a$. At time $t = 0$, its normalized wave function is

$$\psi(x, t = 0) = \sqrt{\frac{8}{5a}}\left[1 + \cos\frac{\pi x}{a}\right]\sin\frac{\pi x}{a}.$$

(a) What results can be obtained from an energy measure?
(b) Write the wave function at a later time t.
(c) Calculate the system energy expectation value at $t = 0$ and t.
(d) Calculate the probability of finding the particle in the left half of the segment at time t.

Solution

For this system, energy eigenvalues and corresponding eigenfunctions are given by:

$$E_n = \frac{\hbar^2 \pi^2 n^2}{2ma^2}, \quad \psi_n(x) = \sqrt{\frac{2}{a}} \sin \frac{n\pi x}{a}, \quad n = 1, 2, 3, \ldots$$

(a) Note that the wave function at $t = 0$ can be rewritten as a linear combination of eigenfunctions of the Hamiltonian:

$$\psi(x, t = 0) = \sqrt{\frac{8}{5a}} \left[\sin \frac{\pi x}{a} + \frac{1}{2} \sin \frac{2\pi x}{a} \right] = \sqrt{\frac{4}{5}} \psi_1(x) + \sqrt{\frac{1}{5}} \psi_2(x).$$

Since the sum of the square modules of the coefficients is equal to 1, the wave function is normalized. It is a superposition of $\psi_1(x)$ and $\psi_2(x)$, thus the values obtained from an energy measure may be only E_1 and E_2.

(b) The wave function at time t is obtained by applying the evolution operator

$$\psi(x, t) = e^{-i\frac{Ht}{\hbar}} \psi(x, t = 0) = \sqrt{\frac{4}{5}} \psi_1(x) e^{-i\frac{\hbar\pi^2}{2ma^2}t} + \sqrt{\frac{1}{5}} \psi_2(x) e^{-i\frac{2\hbar\pi^2}{ma^2}t}.$$

(c) The energy expectation value at $t = 0$ is given by

$$\langle E \rangle_\psi = \frac{4}{5} E_1(x) + \frac{1}{5} E_2(x) = \frac{4}{5} \frac{\hbar^2 \pi^2}{ma^2}.$$

It remains constant in the course of evolution, since the Hamiltonian does not depend on time.

(d) Taking into account the symmetry properties of the wave functions with respect to $x = \frac{a}{2}$, we have

$$P(x < \frac{a}{2}, t) = \int_0^{\frac{a}{2}} dx \, |\psi(x, t)|^2 =$$

$$= \int_0^{\frac{a}{2}} dx \left[\frac{4}{5} |\psi_1(x)|^2 + \frac{1}{5} |\psi_2(x)|^2 + 2\frac{2}{5} \Re \left(\psi_1(x) \psi_2(x) e^{-i\frac{E_1 - E_2}{\hbar}t} \right) \right] =$$

$$= \frac{1}{2} + \frac{4}{5} \frac{2}{a} \cos \frac{E_1 - E_2}{\hbar} t \int_0^{\frac{a}{2}} dx \sin \frac{\pi x}{a} \sin \frac{2\pi x}{a} =$$

$$= \frac{1}{2} + \frac{16}{5\pi} \cos \frac{E_1 - E_2}{\hbar} t \int_0^1 d \sin z \, \sin^2 z = \frac{1}{2} + \frac{16}{15\pi} \cos \frac{3\pi^2 \hbar}{2ma^2} t.$$

5.8 Particle Confined on a Segment (III)

A particle is in a well of infinitely deep potential

$$V(x) = \begin{cases} \infty, & \text{if } x < 0 \text{ and } x > a, \\ 0, & \text{if } 0 < x < a. \end{cases}$$

Assuming that, at time $t = 0$, the particle is in the state described by the wave function

$$\psi(x, 0) = \begin{cases} 0, & \text{see } x < 0 \text{ ex } > a, \\ Ax(a - x), & \text{see } 0 < x < a, \end{cases}$$

determine:

(a) the energy probability distribution;
(b) the energy expectation value and uncertainty;
(c) the wave function at the generic instant t.

Solution

For this system, energy eigenvalues and corresponding eigenfunctions are given by

$$E_n = \frac{\hbar^2 \pi^2 n^2}{2ma^2}, \quad \psi_n(x) = \sqrt{\frac{2}{a}} \sin \frac{n\pi x}{a}, \quad n = 1, 2, 3, \ldots$$

The questions require that the wave function be correctly normalized, so we determine A:

$$\frac{1}{|A|^2} = \int_0^a x^2 (a - x)^2 dx = a^5 \int_0^1 t^2 (1 - t)^2 dt = \frac{a^5}{30} \quad \Rightarrow \quad A = \sqrt{\frac{30}{a^5}},$$

apart from an arbitrary phase factor.

(a) The probability of finding the particle in the nth state is given by the square modulus of

$$c_n = \langle n | \psi \rangle = \int_0^a \sqrt{\frac{2}{a}} \sin \frac{n\pi x}{a} \sqrt{\frac{30}{a^5}} x(a - x) \, dx = \frac{4\sqrt{15}}{n^3 \pi^3} (1 - \cos n\pi) =$$

$$= \frac{4\sqrt{15}}{n^3 \pi^3} (1 - (-1)^n).$$

Since the wave function is symmetric with respect to $x = a/2$, it only has components that have the same property, the eigenfunctions with odd n.

(b) The energy expectation value is given by

$$\langle E \rangle = \langle \psi | \mathcal{H} | \psi \rangle = \int_0^a \psi^*(x) \left(-\frac{\hbar^2}{2m} \frac{d^2}{dx^2} \right) \psi(x) dx =$$

$$= -\frac{\hbar^2}{2m} \frac{30}{a^5} \int_0^a x(a-x)(-2) \, dx =$$

$$= \frac{\hbar^2}{2m} \frac{60}{a^2} \int_0^1 t(1-t)dt = \frac{\hbar^2}{2m} \frac{10}{a^2}.$$

To get the energy uncertainty, we first calculate the expectation value of E^2:

$$\langle E^2 \rangle = \langle \psi | \mathcal{H}\mathcal{H} | \psi \rangle = \int_0^a \left| -\frac{\hbar^2}{2m} \frac{d^2}{dx^2} \psi(x) \right|^2 dx =$$

$$= \frac{\hbar^4}{4m^2} \frac{30}{a^5} 4a = \frac{\hbar^4}{m^2} \frac{30}{a^4}.$$

Then, we obtain

$$\Delta E = \sqrt{\langle E^2 \rangle - \langle E \rangle^2} = \sqrt{\frac{\hbar^4}{m^2 a^4}(30-25)} = \sqrt{5} \frac{\hbar^2}{ma^2}.$$

(c) To know the wave function at instant t, we apply the evolution operator:

$$\psi(x,t) = e^{-i\frac{\mathcal{H}t}{\hbar}} \psi(x, t=0) =$$

$$= \sum_n c_n \psi_n(x) e^{-i\frac{E_n t}{\hbar}} =$$

$$= \sum_n \sqrt{\frac{30}{a}} \frac{8}{(2n+1)^3 \pi^3} \sin\frac{(2n+1)\pi x}{a} \exp\{-i\frac{\hbar\pi^2(2n+1)^2}{2ma^2} t\}.$$

5.9   Harmonic Oscillator (I)

A particle of mass m subjected to a harmonic potential

$$V(x) = \frac{1}{2}m\omega^2 x^2$$

is, at time $t = 0$, in a state determined by the following conditions:

(a) every energy measure gives certain values that satisfy the relationship

$$\hbar\omega < E < 3\hbar\omega;$$

(b) the energy expectation value is

$$\langle E \rangle = \frac{11}{6} \hbar\omega;$$

(c) the position expectation value is:

$$\langle x \rangle = -\sqrt{\frac{8\hbar}{9m\omega}}.$$

Identify this status. Finally, determine the instants at which the position expectation value is positive and maximum.

Solution

The eigenvalue equation for the harmonic oscillator Hamiltonian has the solutions

$$E_n = (n + \frac{1}{2})\hbar\omega \quad ; \quad \mathcal{H}|n\rangle = E_n|n\rangle.$$

Let us now impose the assigned conditions:

(a) The possible measured energy values are

$$E_1 = \frac{3}{2}\hbar\omega \quad \text{and} \quad E_2 = \frac{5}{2}\hbar\omega.$$

The system state vector can therefore be written as

$$|\psi\rangle = c_1|1\rangle + c_2|2\rangle \quad \text{with} \quad |c_1|^2 + |c_2|^2 = 1.$$

(b) The energy expectation value is given by:

$$\langle E \rangle_\psi = |c_1|^2 E_1 + |c_2|^2 E_2 = \frac{1}{2}\hbar\omega(3|c_1|^2 + 5|c_2|^2) = \frac{11}{6}\hbar\omega.$$

This condition, together with the previous one, implies that

$$|c_1|^2 = \frac{2}{3} \quad \text{and} \quad |c_2|^2 = \frac{1}{3}.$$

The phase of one of the coefficients can be arbitrarily set, and therefore

$$c_1 = \sqrt{\frac{2}{3}} \quad \text{and} \quad c_2 = \frac{1}{\sqrt{3}}e^{i\delta}.$$

(c) Using the known properties of the creation and destruction operators (A.15) given in the appendix, we have

$$\langle x \rangle_\psi = (c_1^* \langle 1| + c_2^* \langle 2|) \sqrt{\frac{\hbar}{2m\omega}} (a + a^+)(c_1|1\rangle + c_2|2\rangle) =$$
$$= \sqrt{\frac{\hbar}{2m\omega}} (c_1^* \langle 1| + c_2^* \langle 2|)(\sqrt{2} c_2 |1\rangle + \sqrt{2} c_1 |2\rangle) =$$
$$= \sqrt{\frac{\hbar}{m\omega}} (c_1^* c_2 + c_2^* c_1) = -\sqrt{\frac{8\hbar}{9m\omega}},$$

from which we derive

$$c_1^* c_2 + c_2^* c_1 = -\frac{2\sqrt{2}}{3}.$$

By replacing the expressions for c_1 and c_2, we get:

$$\cos \delta = -1 \quad \Rightarrow \quad \delta = \pi \quad \Rightarrow \quad \begin{cases} c_1 = \sqrt{\frac{2}{3}} \\ c_2 = -\frac{1}{\sqrt{3}} \end{cases}$$

Finally, we get the harmonic oscillator state at $t = 0$:

$$|\psi\rangle = \sqrt{\frac{2}{3}}|1\rangle - \frac{1}{\sqrt{3}}|2\rangle.$$

At time t, by applying the propagator to $|\psi\rangle$,

$$|\psi(t)\rangle = e^{-i\frac{\mathcal{H}t}{\hbar}}|\psi(t)\rangle = \sqrt{\frac{2}{3}} e^{-i\frac{3}{2}\omega t}|1\rangle - \frac{1}{\sqrt{3}} e^{-i\frac{5}{2}\omega t}|2\rangle.$$

Finally, we determine the expectation value of x at time t:

$$\langle x(t) \rangle_\psi = \sqrt{\frac{\hbar}{m\omega}} (c_1^*(t)c_2(t) + c_2^*(t)c_1(t)) =$$
$$= -\frac{4}{3}\sqrt{\frac{\hbar}{2m\omega}} \cos \omega t.$$

Therefore,

$$\langle x(t) \rangle_\psi > 0 \quad \text{if} \quad t \in \left(\frac{(4n+1)\pi}{2\omega}, \frac{(4n+3)\pi}{2\omega} \right) \quad \text{with} \quad n = 0, 1, 2, \ldots,$$
$$\langle x(t) \rangle_\psi = \max_t \langle x(t) \rangle_\psi \quad \text{if} \quad t = \frac{(2n+1)\pi}{\omega} \quad \text{with} \quad n = 0, 1, 2, \ldots.$$

5.10 Harmonic Oscillator (II)

A particle of mass m moves into a harmonic potential with frequency ω. Its state at time $t = 0$ is described by the wave function

$$\psi(x, 0) = A(x^2 + 2\sqrt{\frac{\hbar}{m\omega}}x)e^{-\frac{m\omega x^2}{2\hbar}}.$$

Determine the expression of the wave function at a later time $t > 0$ and the energy expectation value.

Solution

We notice that the wave function is the product of the Gaussian term common to all of the eigenfunctions of the harmonic oscillator for a second-degree polynomial. So, it must be a linear combination of the first three eigenfunctions:

$$\psi(x, 0) = c_0\phi_0(x) + c_1\phi_1(x) + c_2\phi_2(x).$$

They are (A.16)

$$\phi_0(x) = \left(\frac{m\omega}{\pi\hbar}\right)^{\frac{1}{4}} e^{-\frac{m\omega x^2}{2\hbar}},$$

$$\phi_1(x) = \left(\frac{m\omega}{\pi\hbar}\right)^{\frac{1}{4}} \sqrt{2}\sqrt{\frac{m\omega}{\hbar}}xe^{-\frac{m\omega x^2}{2\hbar}},$$

$$\phi_2(x) = \left(\frac{m\omega}{\pi\hbar}\right)^{\frac{1}{4}} \frac{1}{\sqrt{2}}\left(\frac{2m\omega}{\hbar}x^2 - 1\right)e^{-\frac{m\omega x^2}{2\hbar}}.$$

It must therefore result that

$$A(x^2 + 2\sqrt{\frac{\hbar}{m\omega}}x) = \left(\frac{m\omega}{\pi\hbar}\right)^{\frac{1}{4}}\left[c_0 + c_1\sqrt{2}\sqrt{\frac{m\omega}{\hbar}}x + c_2\frac{1}{\sqrt{2}}\left(\frac{2m\omega}{\hbar}x^2 - 1\right)\right].$$

In order for the two sides to coincide, the coefficients of the same powers must be equal, so

$$c_0 = A\left(\frac{m\omega}{\pi\hbar}\right)^{-\frac{1}{4}}\frac{\hbar}{2m\omega},$$

$$c_1 = A\left(\frac{m\omega}{\pi\hbar}\right)^{-\frac{1}{4}}\frac{\hbar\sqrt{2}}{m\omega},$$

$$c_2 = A\left(\frac{m\omega}{\pi\hbar}\right)^{-\frac{1}{4}}\frac{\hbar}{\sqrt{2}m\omega}.$$

We determine A so that ψ is normalized. Since ϕ_n are already normalized, it must result that

$$\sum_{n=0}^{2} |c_n|^2 = 1,$$

from which we easily obtain, setting the arbitrary phase of A equal to zero,

$$A = \sqrt{\frac{4}{11}} \left(\frac{m\omega}{\pi\hbar}\right)^{\frac{1}{4}} \frac{m\omega}{\hbar}$$

and, consequently,

$$c_0 = \frac{1}{\sqrt{11}}, \quad c_1 = \sqrt{\frac{8}{11}}, \quad c_2 = \sqrt{\frac{2}{11}}.$$

Therefore, at time t, the wave function will be

$$\psi(x, t) = \frac{1}{\sqrt{11}} \phi_0(x)e^{-\imath\frac{1}{2}\omega t} + \sqrt{\frac{8}{11}} \phi_1(x)e^{-\imath\frac{3}{2}\omega t} + \sqrt{\frac{2}{11}} \phi_2(x)e^{-\imath\frac{5}{2}\omega t}.$$

The energy expectation value, since the Hamiltonian does not depend on time, is constant and equal to

$$\langle E \rangle = |c_0|^2 E_0 + |c_1|^2 E_1 + |c_2|^2 E_2 = \frac{1}{11}\frac{1}{2}\hbar\omega + \frac{8}{11}\frac{3}{2}\hbar\omega + \frac{2}{11}\frac{5}{2}\hbar\omega = \frac{35}{22}\hbar\omega.$$

5.11 Harmonic Oscillator (III)

A harmonic oscillator of mass m and frequency ω is in such a state that the energy expectation value is

$$\langle \mathcal{H} \rangle = \frac{3}{2}\hbar\omega,$$

the squared uncertainty is given by

$$\langle (\mathcal{H} - \langle \mathcal{H} \rangle)^2 \rangle = \frac{1}{2}(\hbar\omega)^2,$$

and, moreover, an energy measurement cannot give a result greater than $3\hbar\omega$.

(a) What results can be obtained by measuring energy, and with what probability?
(b) Write the most general state vector compatible with the aforementioned information.

(c) Knowing that, at the instant $t = 0$, the expectation value of the position operator is the maximum possible, determine its value at a subsequent instant t.

Solution

(a) As $E \leq 3\hbar\omega$, the results of an energy measurement can be

$$E_0 = \frac{1}{2}\hbar\omega, \quad E_1 = \frac{3}{2}\hbar\omega, \quad E_2 = \frac{5}{2}\hbar\omega,$$

eigenvalues relative to the first three energy eigenstates. The oscillator's status vector is therefore

$$|\psi\rangle = a|0\rangle + b|1\rangle + c|2\rangle, \quad \text{with} \quad |a|^2 + |b|^2 + |c|^2 = 1.$$

Coefficients a, b and c are also subject to the conditions imposed

$$\langle\mathcal{H}\rangle = \left[\frac{1}{2}|a|^2 + \frac{3}{2}|b|^2 + \frac{5}{2}|c|^2\right]\hbar\omega = \frac{3}{2}\hbar\omega$$

and

$$\langle(\mathcal{H} - \langle\mathcal{H}\rangle)^2\rangle = \langle\mathcal{H}^2\rangle - \langle\mathcal{H}\rangle^2 =$$
$$= \left[\frac{1}{4}|a|^2 + \frac{9}{4}|b|^2 + \frac{25}{4}|c|^2\right]\hbar^2\omega^2 - \frac{9}{4}\hbar^2\omega^2 =$$
$$= \frac{1}{2}\hbar^2\omega^2.$$

These three conditions make it possible to determine the square modules of the three coefficients and, therefore, the desired probabilities:

$$P(E = \frac{1}{2}\hbar\omega) = |a|^2 = \frac{1}{4}, \quad P(E = \frac{3}{2}\hbar\omega) = |b|^2 = \frac{1}{2}, \quad P(E = \frac{5}{2}\hbar\omega) = |c|^2 = \frac{1}{4}.$$

(b) If we set the a phase for 0, we can write

$$|\psi\rangle = \frac{1}{2}|0\rangle + \frac{1}{\sqrt{2}}e^{i\beta}|1\rangle + \frac{1}{2}e^{i\gamma}|2\rangle.$$

(c) Using formulas (A.14, A.15), we deduce that

$$X|\psi\rangle = \sqrt{\frac{\hbar}{2m\omega}} \left[b|0\rangle + (a + \sqrt{2}c)|1\rangle + \sqrt{2}b|2\rangle + \sqrt{3}c)|3\rangle \right],$$

$$\langle X \rangle = \langle \psi | X | \psi \rangle =$$

$$= \sqrt{\frac{\hbar}{2m\omega}} \left[a^*b + b^*(a + \sqrt{2}c) + \sqrt{2}c^*b \right] =$$

$$= \sqrt{\frac{\hbar}{2m\omega}} \left[2\Re(a^*b) + 2\sqrt{2}\Re(b^*c) \right] =$$

$$= \sqrt{\frac{\hbar}{2m\omega}} \left[\frac{1}{\sqrt{2}} \cos\beta + \cos(\gamma - \beta) \right].$$

As β and γ are independent, $\langle X \rangle$ takes the maximum value

$$\langle X \rangle_{max} = \frac{2 + \sqrt{2}}{2} \sqrt{\frac{\hbar}{2m\omega}}$$

for

$$\cos\beta = \cos(\gamma - \beta) = 1, \quad \text{that is, for} \quad \beta = \gamma = 0.$$

The desired state is therefore:

$$|\psi\rangle = \frac{1}{2}|0\rangle + \frac{1}{\sqrt{2}}|1\rangle + \frac{1}{2}|2\rangle.$$

At time $t > 0$, it will become

$$|\psi(t)\rangle = \frac{1}{2}e^{-i\frac{1}{2}\omega t}|0\rangle + \frac{1}{\sqrt{2}}e^{-i\frac{3}{2}\omega t}|1\rangle + \frac{1}{2}e^{-i\frac{5}{2}\omega t}|2\rangle.$$

Repeating the calculation already done for $\langle X \rangle$ at $t = 0$, we obtain

$$\langle X(t) \rangle = \frac{2 + \sqrt{2}}{2} \sqrt{\frac{\hbar}{2m\omega}} \cos\omega t.$$

5.12 Plane Rotator

A plane rotator is a rigid two-particle system that can rotate freely in a plane. Denoting its moment of inertia with I, determine:

(a) the system's energy eigenvalues and eigenfunctions;

(b) the possible values of the angular momentum, their probabilities and the angular momentum expectation value in the state described by the wave function

$$\psi(\varphi) = N \cos^2 \varphi;$$

(c) the time evolution of state ψ.

Solution

(a) The system's Hamiltonian is

$$\mathcal{H} = \frac{L^2}{2I} = \frac{L_z^2}{2I} = -\frac{\hbar^2}{2I} \frac{\partial^2}{\partial \varphi^2},$$

where we assumed that the rotation plane is the xy plane. The eigenvalues and eigenfunctions of $\mathcal{H}$ are

$$E_m = \frac{\hbar^2 m^2}{2I}, \quad \psi_m(\varphi) = \frac{1}{\sqrt{2\pi}} e^{im\varphi} \quad \text{with } m = 0, \pm 1, \pm 2, \ldots$$

All of the eigenvalues are doubly degenerate for every m, except for $m = 0$.

(b) We fix the constant N by normalization:

$$1 = |N|^2 \int_0^{2\pi} d\varphi \cos^4 \varphi = |N|^2 \int_0^{2\pi} d\varphi \left[\frac{3}{8} + \frac{1}{2} \cos 2\varphi + \frac{1}{8} \cos 4\varphi \right] =$$

$$= |N|^2 \left[\frac{3}{8} 2\pi + 0 + 0 \right] \quad \Rightarrow \quad |N|^2 = \frac{4}{3\pi} \quad \Rightarrow \quad N = \frac{2}{\sqrt{3\pi}},$$

apart from an arbitrary phase factor. The wave function is a superposition of two Hamiltonian's eigenfunctions. Indeed,

$$\psi(\varphi) = \frac{2}{\sqrt{3\pi}} \cos^2 \varphi = \frac{1}{2\sqrt{3\pi}} \left(e^{2i\varphi} + e^{-2i\varphi} + 2 \right) =$$

$$= \frac{1}{\sqrt{6}} \psi_2(\varphi) + \frac{1}{\sqrt{6}} \psi_{-2}(\varphi) + \sqrt{\frac{2}{3}} \psi_0(\varphi).$$

Taking into account that the $\mathcal{H}$ eigenfunctions are also L_z eigenfunctions, the required probabilities are given by

$$P(L_z = 0) = \frac{2}{3}, \quad P(L_z = 2) = \frac{1}{6}, \quad P(L_z = -2) = \frac{1}{6}.$$

So, the angular momentum expectation value is

$$\langle L_z \rangle = \frac{2}{3} \cdot 0 + \frac{1}{6} \cdot 2 + \frac{1}{6} \cdot (-2) = 0.$$

(c) The rotator state at time $t > 0$ is described by the wave function

$$
\psi(\varphi, t) = \frac{1}{\sqrt{6}}\,\psi_2(\varphi)\,e^{-i\frac{E_2 t}{\hbar}} + \frac{1}{\sqrt{6}}\,\psi_{-2}(\varphi)\,e^{-i\frac{E_2 t}{\hbar}} + \sqrt{\frac{2}{3}}\,\psi_0(\varphi)\,e^{-i\frac{E_0 t}{\hbar}} =
$$

$$
= \frac{1}{\sqrt{3\pi}}\left(\cos 2\varphi\,e^{-i\frac{E_2 t}{\hbar}} + 1\right).
$$

5.13   Rotator in Magnetic Field (I)

The Hamiltonian of an isotropic rotator in the presence of a uniform magnetic field is given by

$$
H = \frac{L^2}{2I} + \alpha L_z,
$$

where α is a constant. Its state is described at time $t = 0$ by the wave function

$$
\psi(\vartheta, \phi) = \sqrt{\frac{15}{32\pi}}\,\sin^2\theta\,\sin 2\phi.
$$

Determine the expression of the wave function at a later time t.

Solution

The Hamiltonian's eigenfunctions are the Spherical Harmonics and the eigenvalues are

$$
\frac{\ell(\ell+1)\hbar^2}{2I} + \alpha\hbar m, \quad \ell = 0, 1, 2, \ldots, \quad m = 0, \pm 1, \ldots
$$

The wave function that represents the initial state can be written, using (A.40), as:

$$
\psi(\vartheta, \phi, t = 0) = \frac{1}{2i}\sqrt{\frac{15}{32\pi}}\,\sin^2(\theta)\left(e^{2i\phi} - e^{-2i\phi}\right) = \frac{1}{2i}\left(Y_2^2(\vartheta, \phi) - Y_2^{-2}(\vartheta, \phi)\right).
$$

At a later time t, we will have

$$
\psi(\vartheta, \phi, t) = e^{-i\frac{\mathcal{H}t}{\hbar}}\psi((\vartheta, \phi), t = 0) =
$$

$$
= \frac{1}{2i}\left(Y_2^2(\vartheta, \phi)e^{-i[\frac{6\hbar}{2I} + 2\alpha]t} - Y_2^{-2}(\vartheta, \phi)e^{-i[\frac{6\hbar}{2I} - 2\alpha]t}\right).
$$

5.14   Rotator in Magnetic Field (II)

The Hamiltonian of an isotropic rotator in the presence of a uniform magnetic field is given by

$$\mathcal{H} = \frac{L^2}{2I} + \alpha L_z.$$

At instant $t = 0$, its wave function is

$$\psi(\vartheta, \phi, 0) = \frac{1}{\sqrt{2}} \left[Y_1^{+1}(\vartheta, \phi) + Y_1^{-1}(\vartheta, \phi) \right].$$

(a) What is its wave function at time t?
(b) Determine the instant t at which the wave function is proportional to

$$\frac{1}{\sqrt{2}} \left(Y_1^{+1}(\vartheta, \phi) - Y_1^{-1}(\vartheta, \phi) \right).$$

Solution

The system is the same as in problem 5.13.

(a) At any instant $t > 0$, we will have

$$\psi(\vartheta, \phi, t) = e^{-i\frac{\mathcal{H}t}{\hbar}} \psi(\vartheta, \phi, t = 0) =$$
$$= \frac{1}{\sqrt{2}} \left[Y_1^{+1}(\vartheta, \phi) e^{-i[\frac{\hbar}{I}+\alpha]t} + Y_1^{-1}(\vartheta, \phi) e^{-i[\frac{\hbar}{I}-\alpha]t} \right].$$

(b) To answer the second question, it is sufficient to impose that the ratio between the two Spherical Harmonics coefficients is equal to -1:

$$\frac{e^{-i[\frac{\hbar}{I}-\alpha]t}}{e^{-i[\frac{\hbar}{I}+\alpha]t}} = e^{-2i\alpha t} = -1 \quad \Rightarrow \quad 2\alpha t = (2n+1)\pi \quad \Rightarrow \quad t = \frac{(2n+1)\pi}{2\alpha}.$$

5.15   Fermion in a Magnetic Field (I)

At time $t = 0$, a particle of spin $\frac{1}{2}$, magnetic moment $\mu = g\,\mathbf{S}$ and infinite mass is in a state in which the probability of finding the value $\hbar/2$ making a measure of S_z is 2/3, and the expectation values of S_x and S_y are equal and both are positive. The particle is immersed in a constant magnetic field $\mathbf{B}$ parallel to the y axis.

(a) Write the state vector at time $t = 0$ and determine the common expectation value of S_x and S_y.
(b) Calculate the maximum and minimum values of the probability of finding the value $\hbar/2$ in a measure of S_z during the time evolution of the system.

Solution

(a) Due to the infinite mass, such a particle has no kinetic energy, so the system
    Hamiltonian is

$$\mathcal{H} = -g\mathbf{S} \cdot \mathbf{B} = -gBS_y.$$

Apart from an arbitrary overall phase factor, the state of the system at the initial
instant can be written in terms of S_z eigenstates in the following form:

$$|\psi(0)\rangle = \sqrt{\frac{2}{3}}|+\rangle + e^{i\alpha}\sqrt{\frac{1}{3}}|-\rangle = \sqrt{\frac{1}{3}}\begin{pmatrix} \sqrt{2} \\ e^{i\alpha} \end{pmatrix}.$$

The expectation values of S_x and S_y are given by

$$\frac{2}{\hbar}\langle S_x\rangle = \frac{1}{3}\left(\sqrt{2}\,e^{-i\alpha}\right)\begin{pmatrix} 0 & 1 \\ 1 & 0 \end{pmatrix}\begin{pmatrix} \sqrt{2} \\ e^{i\alpha} \end{pmatrix} = \frac{\sqrt{2}}{3}\left(e^{i\alpha} + e^{-i\alpha}\right) = \frac{2\sqrt{2}}{3}\cos\alpha,$$

$$\frac{2}{\hbar}\langle S_y\rangle = \frac{1}{3}\left(\sqrt{2}\,e^{-i\alpha}\right)\begin{pmatrix} 0 & -i \\ i & 0 \end{pmatrix}\begin{pmatrix} \sqrt{2} \\ e^{i\alpha} \end{pmatrix} = i\frac{\sqrt{2}}{3}\left(-e^{i\alpha} + e^{-i\alpha}\right) = \frac{2\sqrt{2}}{3}\sin\alpha.$$

Imposing that they are equal, we get

$$\langle S_x\rangle = \langle S_y\rangle \quad \Rightarrow \quad \sin\alpha = \cos\alpha \quad \Rightarrow \alpha = \frac{\pi}{4},$$

because $\langle S_x\rangle$ and $\langle S_y\rangle$ must be positive. Ultimately, the state at $t = 0$ is

$$|\psi(0)\rangle = \sqrt{\frac{2}{3}}\left(|+\rangle + \frac{1+i}{2}|-\rangle\right)$$

and the common expectation value of S_x and S_y is

$$\langle S_x\rangle = \langle S_y\rangle = \frac{\hbar}{2}\frac{2\sqrt{2}}{3}\cos\frac{\pi}{4} = \frac{\hbar}{3}.$$

(b) At time $t > 0$,

$$|\psi(t)\rangle = e^{-i\frac{\mathcal{H}t}{\hbar}}|\psi(0)\rangle = e^{i\omega\sigma_y t}\sqrt{\frac{2}{3}}\begin{pmatrix} 1 \\ \frac{1+i}{2} \end{pmatrix},$$

where we introduced the quantity $\omega = gB/2$. Applying the well-known property
of Pauli matrices (A.74), we obtain

$$|\psi(t)\rangle = \left(I \cos \omega t + i\sigma_y \sin \omega t\right) \sqrt{\frac{2}{3}} \begin{pmatrix} 1 \\ \frac{1+i}{2} \end{pmatrix} =$$

$$= \sqrt{\frac{2}{3}} \begin{pmatrix} \cos \omega t & \sin \omega t \\ -\sin \omega t & \cos \omega t \end{pmatrix} \begin{pmatrix} 1 \\ \frac{1+i}{2} \end{pmatrix} =$$

$$= \sqrt{\frac{2}{3}} \begin{pmatrix} \cos \omega t + \frac{1+i}{2} \sin \omega t \\ \frac{1+i}{2} \cos \omega t - \sin \omega t \end{pmatrix}$$

The probability of finding the value $\hbar/2$ in a measure of S_z is given by the square modulus of the pertinent component:

$$P\left(S_z = +\frac{\hbar}{2}\right) = \frac{2}{3} \left| \cos \omega t + \frac{1+i}{2} \sin \omega t \right|^2 =$$

$$= \frac{2}{3} \left(\cos^2 \omega t + \frac{1}{2} \sin^2 \omega t + \sin \omega t \cos \omega t \right).$$

The maximum is obtained by imposing the condition

$$\frac{dP}{dt} = \frac{2}{3} \omega \left(\cos 2\omega t - \frac{\sin 2\omega t}{2} \right) = 0 \quad \Rightarrow \quad \tan 2\omega t = 2.$$

This equation has two solutions. The first one,

$$t = \frac{1}{2\omega} [\arctan 2 + (2n+1)\pi],$$

corresponds to the desired maximum. There is also another solution,

$$t = \frac{1}{2\omega} (\arctan 2 + 2n\pi),$$

but it corresponds to a minimum point, because the second derivative is negative there.

5.16 Fermion in a Magnetic Field (II)

A particle of infinite mass and spin $\frac{1}{2}$ is, at time $t = 0$, in a state in which the probability of finding the spin component along the positive direction of the z axis is $\frac{1}{4}$ and, along the negative direction, $\frac{3}{4}$. This information determines the status apart from a parameter.

The particle is subjected to a constant and uniform magnetic field $\mathbf{B}$ along the x axis.

(a) Write the expression of the initial state (including the indeterminate parameter).
(b) Write the system's Hamiltonian (the particle's magnetic moment operator is $\mu = g\mathbf{S}$).
(c) Write the expression of the state (always containing an indeterminate parameter) as a function of time.
(d) Determine the parameter values for which the wave function at a certain time is reduced to a σ_z eigenstate and find the times at which this happens.

Solution

(a) Apart from an arbitrary overall phase factor, the state of the system at the initial instant can be written in terms of the S_z eigenstates in the form

$$|\psi(0)\rangle = \frac{1}{2}|+\rangle + \frac{\sqrt{3}}{2}e^{i\alpha}|-\rangle = \frac{1}{2}\begin{pmatrix} 1 \\ \sqrt{3}\,e^{i\alpha} \end{pmatrix}.$$

(b) Due to the infinite mass, such a particle has no kinetic energy, so the system Hamiltonian is

$$\mathcal{H} = -g\mathbf{S} \cdot \mathbf{B} = -gBS_x = -\hbar\omega\sigma_x \quad \sigma_x = \begin{pmatrix} 0 & 1 \\ 1 & 0 \end{pmatrix},$$

where

$$\omega = \frac{gB}{2}.$$

The eigenvalues and the corresponding eigenvectors of $\mathcal{H}$ are

$$E_1 = -\hbar\omega \quad |1\rangle = \frac{1}{\sqrt{2}}\begin{pmatrix} 1 \\ 1 \end{pmatrix} = \frac{1}{\sqrt{2}}(|+\rangle + |-\rangle),$$

$$E_2 = \hbar\omega \quad |2\rangle = \frac{1}{\sqrt{2}}\begin{pmatrix} 1 \\ -1 \end{pmatrix} = \frac{1}{\sqrt{2}}(|+\rangle - |-\rangle).$$

(c) At times $t > 0$,

$$|\psi(t)\rangle = e^{-i\frac{\mathcal{H}t}{\hbar}}|\psi(0)\rangle = e^{i\omega t}|1\rangle\langle 1|\psi(0)\rangle + e^{-i\omega t}|2\rangle\langle 2|\psi(0)\rangle =$$

$$= \frac{e^{i\omega t}}{2\sqrt{2}}\left(1 + \sqrt{3}e^{i\alpha}\right)\frac{1}{\sqrt{2}}(|+\rangle + |-\rangle) +$$

$$+ \frac{e^{-i\omega t}}{2\sqrt{2}}\left(1 - \sqrt{3}e^{i\alpha}\right)\frac{1}{\sqrt{2}}(|+\rangle - |-\rangle) =$$

$$= \frac{1}{2}\left[\cos\omega t + i\sqrt{3}e^{i\alpha}\sin\omega t\right]|+\rangle + \frac{1}{2}\left[i\sin\omega t + \sqrt{3}e^{i\alpha}\cos\omega t\right]|-\rangle.$$

(d) Suppose that, at the time $t = T$, it happens that $|\psi(T)\rangle = |+\rangle$. It must result that

$$i \sin \omega T + \sqrt{3}e^{i\alpha} \cos \omega T = 0,$$
$$\sin \omega T - \sqrt{3}e^{i(\alpha+\pi/2)} \cos \omega T = 0,$$
$$\tan \omega T = \sqrt{3}e^{i(\alpha+\pi/2)}.$$

Imposing that ωT is real, we obtain

$$\alpha + \pi/2 = 0, \pi \quad \Rightarrow \quad \alpha = -\pi/2, \pi/2.$$

In the two cases, we have

$$\alpha = -\pi/2 \quad \Rightarrow \quad T = \frac{1}{\omega} \arctan \sqrt{3} = \frac{\pi}{\omega}\left(n + \frac{1}{3}\right), \quad n = 0, 1, \ldots,$$

$$\alpha = \pi/2 \quad \Rightarrow \quad T = \frac{1}{\omega} \arctan(-\sqrt{3}) = \frac{\pi}{\omega}\left(n - \frac{1}{3}\right), \quad n = 1, 2, \ldots,$$

where it has been taken into account that $T > 0$.

5.17 Fermion in a Magnetic Field (III)

The Hamiltonian of a spin $\frac{1}{2}$ particle is

$$\mathcal{H} = -g\mathbf{S} \cdot \mathbf{B},$$

where $\mathbf{S}$ is the spin and $\mathbf{B}$ is a magnetic field directed along the z axis. Determine:

(a) The explicit form as a function of $\mathbf{S}$ and $\mathbf{B}$ of the operator $\dot{\mathbf{S}}$.
(b) The eigenstates of $(\dot{\mathbf{S}})_y$ and the relative eigenvalues.
(c) The time evolution of a state that coincides at time $t = 0$ with one of the afore-mentioned eigenstates, and the energy expectation value.

Solution

(a) Heisenberg's equation for operator $\mathbf{S}$ is

$$\dot{\mathbf{S}} = \frac{d\mathbf{S}}{dt} = \frac{i}{\hbar}[\mathcal{H}, \mathbf{S}] = -\frac{igB}{\hbar}[S_z, \mathbf{S}]$$

(as $\mathbf{S}$ does not depend explicitly on time, $\frac{\partial \mathbf{S}}{\partial t} = 0$). Component by component,

$$\dot{S}_x = -\frac{igB}{\hbar}[S_z, S_x] = gBS_y,$$

$$\dot{S}_y = -\frac{igB}{\hbar}[S_z, S_y] = -gBS_x,$$

$$\dot{S}_z = -\frac{igB}{\hbar}[S_z, S_z] = 0,$$

where we used $\mathbf{S} = \frac{\hbar}{2}\boldsymbol{\sigma}$ and the Pauli matrices commutation relations (A.71).

(b) The eigenvalues of $(\dot{\mathbf{S}})_y$ are the eigenvalues of $-\frac{gB\hbar}{2}\sigma_x$, that is,

$$\lambda_1 = -\hbar\omega, \quad \lambda_2 = \hbar\omega, \quad \text{where} \quad \omega = \frac{gB}{2}.$$

The eigenstates are the same as those of σ_x, that is, in σ_z representation,

$$|\lambda_1\rangle = \frac{1}{\sqrt{2}}\begin{pmatrix} 1 \\ 1 \end{pmatrix}, \quad |\lambda_2\rangle = \frac{1}{\sqrt{2}}\begin{pmatrix} 1 \\ -1 \end{pmatrix}.$$

(c) Let us suppose that, at time $t = 0$, the system is in the state

$$|\psi(0)\rangle = \frac{1}{\sqrt{2}}\begin{pmatrix} 1 \\ 1 \end{pmatrix} = \frac{1}{\sqrt{2}}\left[\begin{pmatrix} 1 \\ 0 \end{pmatrix} + \begin{pmatrix} 0 \\ 1 \end{pmatrix}\right].$$

At time $t > 0$, it will be in the state

$$|\psi(t)\rangle = e^{-i\frac{\mathcal{H}t}{\hbar}}|\psi(0)\rangle = \frac{1}{\sqrt{2}}\left[e^{-i\omega t}\begin{pmatrix} 1 \\ 0 \end{pmatrix} + e^{i\omega t}\begin{pmatrix} 0 \\ 1 \end{pmatrix}\right] = \frac{1}{\sqrt{2}}\begin{pmatrix} e^{-i\omega t} \\ e^{i\omega t} \end{pmatrix}.$$

The energy expectation value, since the Hamiltonian does not depend on time, is constant and can be calculated at $t = 0$:

$$\langle E \rangle_\psi = -gB\langle\psi(0)|S_z|\psi(0)\rangle =$$

$$= -\frac{gB}{2}\frac{\hbar}{2}(1\ 1)\begin{pmatrix} 1 & 0 \\ 0 & -1 \end{pmatrix}\begin{pmatrix} 1 \\ 1 \end{pmatrix} =$$

$$= -\frac{gB\hbar}{4}(1\ 1)\begin{pmatrix} 1 \\ -1 \end{pmatrix} = 0.$$

5.18   Fermion in a Magnetic Field (IV)

A particle of infinite mass, spin $\frac{1}{2}$ and magnetic moment $\mu = g\mathbf{S}$, where $\mathbf{S}$ is the spin, is placed in a constant magnetic field directed along the x axis. At time $t = 0$, the spin component $S_z = \frac{\hbar}{2}$ is measured. Find the probability that the particle is found to have $S_y = \pm\frac{\hbar}{2}$ at any subsequent moment.

Solution

The Hamiltonian of the system, neglecting the kinetic term, is

$$\mathcal{H} = -g\mathbf{S} \cdot \mathbf{B} = -\frac{gB\hbar}{2}\sigma_x = -\hbar\omega\sigma_x, \quad \text{where} \quad \omega = \frac{gB}{2}.$$

We choose to solve the problem using the Schrödinger equation that controls the evolution of quantum states. In the S_z representation, this equation is

$$i\hbar\frac{d}{dt}\begin{pmatrix} a_1 \\ a_2 \end{pmatrix} + \hbar\omega\begin{pmatrix} 0 & 1 \\ 1 & 0 \end{pmatrix}\begin{pmatrix} a_1 \\ a_2 \end{pmatrix} = 0.$$

Splitting the spin components, we get

$$\begin{cases} i\frac{da_1}{dt} + \omega a_2 = 0 \\ i\frac{da_2}{dt} + \omega a_1 = 0 \end{cases} \quad \Rightarrow \quad \frac{d^2a_{1,2}}{dt^2} + \omega^2 a_{1,2} = 0,$$

the solutions to which are

$$a_{1,2}(t) = A_{1,2}e^{i\omega t} + B_{1,2}e^{-i\omega t}.$$

At $t = 0$, after the measurement, the particle state is the S_z eigenvector relative to the eigenvalue $+\hbar/2$, and therefore

$$a_1(0) = 1 \quad \text{and} \quad a_2(0) = 0 \quad \Leftrightarrow \quad A_1 + B_1 = 1 \quad \text{and} \quad A_2 + B_2 = 0.$$

By imposing these conditions in the Schrödinger equation, we have

$$i\frac{da_1}{dt}\Big|_{t=0} = -\omega A_1 + \omega B_1 = -\omega\, a_2|_{t=0} = 0,$$

$$i\frac{da_2}{dt}\Big|_{t=0} = -\omega A_2 + \omega B_2 = -\omega\, a_1|_{t=0} = -\omega.$$

We have a total of 4 equations:

$$\begin{cases} A_1 + B_1 = 1, \\ A_1 - B_1 = 0, \\ A_2 + B_2 = 0, \\ A_2 - B_2 = 1, \end{cases}$$

the solution to which is

$$A_1 = B_1 = A_2 = -B_2 = \frac{1}{2}.$$

Substituting these values in the $a_{1,2}$ equations, one finds that

$$\psi(t) = \begin{pmatrix} a_1(t) \\ a_2(t) \end{pmatrix} = \begin{pmatrix} \cos \omega t \\ i \sin \omega t \end{pmatrix},$$

which is normalized. S_y eigenstates are easily calculated to be

$$\left| S_y = +\frac{\hbar}{2} \right\rangle = \frac{1}{\sqrt{2}} \begin{pmatrix} 1 \\ i \end{pmatrix}, \quad \left| S_y = -\frac{\hbar}{2} \right\rangle = \frac{1}{\sqrt{2}} \begin{pmatrix} 1 \\ -i \end{pmatrix};$$

so the required probabilities are

$$P(S_y = +\frac{\hbar}{2}) = \left| \langle S_y = +\frac{\hbar}{2} | \psi(t) \rangle \right|^2 = \frac{1}{2} \left| (1 \ -i) \begin{pmatrix} \cos \omega t \\ i \sin \omega t \end{pmatrix} \right|^2 =$$

$$= \frac{1}{2} (\cos \omega t + \sin \omega t)^2 = \frac{1}{2} (1 + \sin 2\omega t),$$

$$P(S_y = -\frac{\hbar}{2}) = \left| \langle S_y = -\frac{\hbar}{2} | \psi(t) \rangle \right|^2 = \frac{1}{2} \left| (1 \ i) \begin{pmatrix} \cos \omega t \\ i \sin \omega t \end{pmatrix} \right|^2 =$$

$$= \frac{1}{2} (\cos \omega t - \sin \omega t)^2 = \frac{1}{2} (1 - \sin 2\omega t).$$

As a control, it can be easily verified that their sum is 1.

5.19   Fermion in a Magnetic Field (V)

At time $t = 0$, a particle of spin $\frac{1}{2}$, magnetic moment $\mu = g\,\mathbf{S}$ and infinite mass is in a state with a z spin component equal to $+\hbar/2$. It is subject to a magnetic interaction of the type

$$\mathcal{H} = \frac{A}{\sqrt{2}} (\sigma_x + \sigma_y),$$

where A is a constant and σ_x and σ_y are Pauli matrices. Calculate how long it takes for the spin component along the z axis to become $-\hbar/2$.

Solution

Having denoted by $\hat{n}$ the unit vector

$$\hat{n} = \left(\frac{1}{\sqrt{2}}, \frac{1}{\sqrt{2}}, 0 \right)$$

and with $\boldsymbol{\vartheta}$ the vector

$$\boldsymbol{\vartheta} = \frac{At}{\hbar} \hat{n},$$

the evolution operator of this system can be written as

$$e^{-i\frac{\mathcal{H}t}{\hbar}} = e^{-i\vartheta \cdot \boldsymbol{\sigma}} = \mathbb{I} \cos \vartheta - i(\mathbf{n} \cdot \boldsymbol{\sigma}) \sin \vartheta,$$

where formula (A.74) has been used. Therefore, the particle status at time t will be given by

$$|\psi(t)\rangle = e^{-i\frac{\mathcal{H}t}{\hbar}} = \left(\mathbb{I} \cos \frac{At}{\hbar} - \frac{i(\sigma_x + \sigma_y)}{\sqrt{2}} \sin \frac{At}{\hbar} \right) \begin{pmatrix} 1 \\ 0 \end{pmatrix} =$$

$$= \cos \frac{At}{\hbar} \begin{pmatrix} 1 \\ 0 \end{pmatrix} - \frac{i}{\sqrt{2}} \sin \frac{At}{\hbar} \begin{pmatrix} 0 \\ 1 + i \end{pmatrix}.$$

In order that, at time t, only the component with $S_z = -\frac{\hbar}{2}$ will survive, it must be $\cos \frac{At}{\hbar} = 0$. This happens for the first time at

$$t = \frac{\pi \hbar}{2A}.$$

5.20 Fermion in a Magnetic Field (VI)

A particle of spin $\frac{1}{2}$, magnetic moment $\mu = g\,\mathbf{S}$ and infinite mass is placed in a uniform and constant magnetic field $\mathbf{B_0}$ directed along the z axis. During the interval $0 < t < T$, a uniform and constant magnetic field $\mathbf{B_1}$ along the x axis is also applied, so that the system will still be in a uniform and constant magnetic field $\mathbf{B} = \mathbf{B_0} + \mathbf{B_1}$. For $t < 0$, the system is in the S_z eigenstate corresponding to the eigenvalue $+\frac{\hbar}{2}$.

(a) At $t = 0^+$, what will be the probability amplitudes of finding the spin component along the $\mathbf{B}$ direction equal to $\pm\frac{\hbar}{2}$?
(b) How do the energy eigenstates evolve over time in the interval $0 < t < T$?
(c) What is the probability of observing the system at the time $t = T$ in the S_z eigenstate corresponding to the eigenvalue $-\frac{\hbar}{2}$?

Solution

For $t < 0$, the system Hamiltonian is

$$\mathcal{H} = -g\mathbf{S} \cdot \mathbf{B_0} = -\frac{g B_0 \hbar}{2} \sigma_z = -\hbar \omega_0 \sigma_z, \quad \text{where} \quad \omega_0 = \frac{g B_0}{2},$$

In the interval between 0 and T, the Hamiltonian becomes

$$\mathcal{H} = -g\mathbf{S} \cdot \mathbf{B} = -\frac{g\hbar}{2}(B_0 \sigma_z + B_1 \sigma_x) = -\frac{g\hbar}{2} \begin{pmatrix} B_0 & B_1 \\ B_1 & -B_0 \end{pmatrix} = -\hbar\omega \begin{pmatrix} \cos \vartheta & \sin \vartheta \\ \sin \vartheta & -\cos \vartheta \end{pmatrix}.$$

$$\text{where}\quad \omega = \frac{gB}{2}, \quad B = \sqrt{B_0^2 + B_1^2}\quad\text{and}\quad \vartheta = \arctan\frac{B_1}{B_0}.$$

This Hamiltonian is proportional to the spin component in the $\hat{n}$ direction, set by the ϑ angle, which is given by

$$\mathbf{S}\cdot\hat{n} = -\frac{\hbar}{2}\begin{pmatrix} \cos\vartheta & \sin\vartheta \\ \sin\vartheta & -\cos\vartheta \end{pmatrix}.$$

$\mathbf{S}\cdot\hat{n}$ has eigenvalues $\pm\frac{\hbar}{2}$ and eigenstates (see Problem 4.2)

$$|\chi_+\rangle = |\mathbf{S}\cdot\hat{n} = +\frac{\hbar}{2}\rangle = \begin{pmatrix} \cos\frac{\vartheta}{2} \\ \sin\frac{\vartheta}{2} \end{pmatrix}, \quad |\chi_-\rangle = |\mathbf{S}\cdot\hat{n} = -\frac{\hbar}{2}\rangle = \begin{pmatrix} -\sin\frac{\vartheta}{2} \\ \cos\frac{\vartheta}{2} \end{pmatrix}.$$

(a) The state of the system at the time $t = 0$ can be expressed in terms of the eigenstates of $\mathbf{S}\cdot\hat{n}$:

$$|\psi(0)\rangle = \begin{pmatrix} 1 \\ 0 \end{pmatrix} = c_+|\chi_+\rangle + c_-|\chi_-\rangle.$$

c_+ e c_- are the amplitudes that we are looking for:

$$c_+ = \langle\chi_+|\psi(0)\rangle = \begin{pmatrix} \cos\frac{\vartheta}{2} & \sin\frac{\vartheta}{2} \end{pmatrix}\begin{pmatrix} 1 \\ 0 \end{pmatrix} = \cos\frac{\vartheta}{2},$$

$$c_- = \langle\chi_-|\psi(0)\rangle = \begin{pmatrix} -\sin\frac{\vartheta}{2} & \cos\frac{\vartheta}{2} \end{pmatrix}\begin{pmatrix} 1 \\ 0 \end{pmatrix} = -\sin\frac{\vartheta}{2}.$$

(b) As we said,

$$\mathcal{H} = -g\mathbf{S}\cdot\mathbf{B} = -gBS\cdot\hat{n},$$

therefore, its eigenkets are $|\chi_+\rangle$ and $|\chi_-\rangle$ and its eigenvalues are

$$E_\pm = -gB\left(\pm\frac{\hbar}{2}\right) = \mp\hbar\omega.$$

In the time interval between $t = 0$ and $t = T$, the eigenstates' evolution is given by

$$|\chi_\pm(t)\rangle = e^{-i\frac{E_\pm t}{\hbar}}|\chi_\pm(0)\rangle = e^{\pm i\omega t}|\chi_\pm(0)\rangle.$$

(c) At time T, the system state is

$$|\psi(T)\rangle = c_+ e^{i\omega T} |\chi_+\rangle + c_- e^{-i\omega T} |\chi_-\rangle =$$

$$= \cos\frac{\vartheta}{2} e^{i\omega T} \begin{pmatrix} \cos\frac{\vartheta}{2} \\ \sin\frac{\vartheta}{2} \end{pmatrix} - \sin\frac{\vartheta}{2} e^{-i\omega T} \begin{pmatrix} -\sin\frac{\vartheta}{2} \\ \cos\frac{\vartheta}{2} \end{pmatrix} =$$

$$= \begin{pmatrix} \cos^2\frac{\vartheta}{2} e^{i\omega T} + \sin^2\frac{\vartheta}{2} e^{-i\omega T} \\ \cos\frac{\vartheta}{2}\sin\frac{\vartheta}{2} e^{i\omega T} - \sin\frac{\vartheta}{2}\cos\frac{\vartheta}{2} e^{-i\omega T} \end{pmatrix}.$$

The required probability is

$$P(S_z = -\hbar/2) = |\langle S_z = -\hbar/2 \,|\psi(T)\rangle|^2 = \left|\cos\frac{\vartheta}{2}\sin\frac{\vartheta}{2}\left(e^{i\omega T} - e^{-i\omega T}\right)\right|^2 =$$

$$= |i\sin\vartheta \, \sin\omega T|^2 = \sin^2\vartheta \, \sin^2\omega T.$$

5.21 Measurements of a Hydrogen Atom

At time $t = 0$, the state of a Hydrogen atom is described by the wave function

$$\psi(\mathbf{r}, t = 0) = \frac{1}{\sqrt{10}}\left(2\psi_{1,0,0} + \psi_{2,1,0} + \sqrt{2}\,\psi_{2,1,1} + \sqrt{3}\,\psi_{2,1,-1}\right),$$

where $\psi_{n,\ell,m}$ are the Hamiltonian eigenfunctions relative to the quantum numbers n, ℓ, m relative to energy, angular momentum square modulus and its z-component. Neglecting spin, determine:

(a) the energy expectation value;
(b) the expression of the wave function at a later time t;
(c) the probability of finding, at time t, the atom in the state with $\ell = 1$, $m = 1$.

Solution

(a) Having called as $c_{n,\ell,m}$ the coefficients of the wave function expansion in the Hamiltonian's eigenfunctions,

$$\psi(\mathbf{r}, t = 0) = \sum_{n,\ell,m} c_{n,\ell,m}\,\psi_{n,\ell,m},$$

since it occurs that

$$\sum_{n,\ell,m} |c_{n,\ell,m}|^2 = \frac{1}{10}(4 + 1 + 2 + 3) = 1,$$

the wave function is normalized. Therefore, the energy expectation value is given by

$$\langle E \rangle = \langle \psi | \mathcal{H} | \psi \rangle = \sum_{n,\ell,m} |c_{n,\ell,m}|^2 E_n = \frac{1}{10} [4E_1 + (1 + 2 + 3)E_2] =$$

$$= \frac{11}{20} E_1 = -\frac{11}{20} \frac{1}{2} m_e c^2 \alpha^2 = -\frac{11 \cdot 0.51}{40 \cdot 137^2} \, \text{eV} = -7.47 \, \text{eV},$$

being that $E_2 = \frac{E_1}{4}$ and $m_e = 0.51 \frac{eV}{c^2}$.

(b) At a later instant t, the wave function will be

$$\psi(\mathbf{r}, t) = e^{-i\frac{\mathcal{H}t}{\hbar}} \psi(\mathbf{r}, t = 0) =$$

$$= \frac{1}{\sqrt{10}} \left(2 \psi_{1,0,0} e^{-i\frac{E_1 t}{\hbar}} + \left[\psi_{2,1,0} + \sqrt{2} \, \psi_{2,1,1} + \sqrt{3} \, \psi_{2,1,-1} \right] e^{-i\frac{E_2 t}{\hbar}} \right).$$

(c) Denoting the required probability with $P_{n,1,1}$, we have

$$P_{n,1,1}(t) = |\langle n, 1, 1 | \psi(t) \rangle|^2 = |\langle n, 1, 1 | e^{-i\frac{\mathcal{H}t}{\hbar}} | \psi(0) \rangle|^2 =$$

$$= \left| \langle n, 1, 1 | \frac{1}{\sqrt{10}} \left(2 | 1, 0, 0 \rangle e^{-i\frac{E_1 t}{\hbar}} + \left[|2, 1, 0 \rangle + \sqrt{2} |2, 1, 1 \rangle + \sqrt{3} |2, 1, -1 \rangle \right] e^{-i\frac{E_2 t}{\hbar}} \right) \right|^2 =$$

$$= \left| \delta_{n,2} \frac{\sqrt{2}}{\sqrt{10}} e^{-i\frac{E_2 t}{\hbar}} \right|^2 = \frac{1}{5} \delta_{n,2}.$$

Chapter 6
Time-Independent Perturbation Theory

6.1 Particle on a Segment: Square Perturbation

A particle of mass m is constrained to move along an L length segment in the presence of a small potential well, so that the total potential is

$$V(x) = \begin{cases} \infty, & \text{if } x < 0 \text{ and } x > L, \\ -V_0, & \text{if } 0 < x < \frac{L}{2}, \\ 0 & \text{if } \frac{L}{2} < x < L. \end{cases}$$

Consider the small potential well (see Fig. 6.1) between 0 and $\frac{L}{2}$ as a perturbation compared to the infinite confining well and calculate the energy eigenvalues at the first perturbative order.

Solution

In the absence of perturbation, the energy eigenvalues and eigenfunctions are given by:

$$E_n^0 = \frac{\hbar^2 \pi^2 n^2}{2mL^2}, \quad \psi_n^0(x) = \sqrt{\frac{2}{L}} \sin \frac{n\pi x}{L}, \quad n = 1, 2, \ldots$$

The first-order correction in the energy levels is given by

$$\begin{aligned} E_n^1 &= \int_0^L dx\, \psi_n^*(x)\, H_1(x)\, \psi_n(x) = \\ &= \frac{2}{L} \int_0^{\frac{L}{2}} dx\, \sin^2 \frac{n\pi x}{L} (-V_0) = \\ &= -\frac{V_0}{L} \int_0^{\frac{L}{2}} dx \left[1 - \cos \frac{2n\pi x}{L} \right] = -\frac{V_0}{2}. \end{aligned}$$

This correction is the same for all levels. We will therefore have

© Springer Nature Switzerland AG 2019
L. Angelini, *Solved Problems in Quantum Mechanics*, UNITEXT for Physics,
https://doi.org/10.1007/978-3-030-18404-9_6

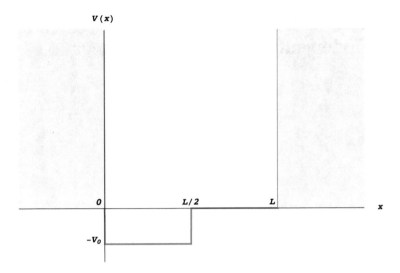

Fig. 6.1 Confining potential with a small well

$$E_n^0 + E_n^1 = \frac{\hbar^2 \pi^2 n^2}{2mL^2} - \frac{V_0}{2}, \qquad \forall n = 1, 2, \ldots$$

6.2   Particle on a Segment: Linear Perturbation

Using first-order Perturbation Theory, calculate the energy levels for a one-dimensional infinite square well of width L, whose bottom was made oblique, as shown in the Fig. 6.2.

Solution

The perturbative potential is given by

$$\mathcal{H}_1 = \frac{V_0}{L} x \qquad \text{for } 0 \leq x \leq L.$$

Unperturbed Hamiltonian eigenvalues and eigenfunctions are

$$E_n^0 = \frac{\hbar^2 \pi^2 n^2}{2mL^2}, \qquad \psi_n^0(x) = \sqrt{\frac{2}{L}} \sin \frac{n\pi x}{L}, \qquad n = 1, 2, \ldots$$

The first-order shift in energy is given by

$$E_n^1 = \frac{V_0}{L} \frac{2}{L} \int_0^L dx\, x\, \sin^2 \frac{n\pi x}{L} = \frac{2V_0}{n^2\pi^2} \int_0^{n\pi} dz\, z\, \sin^2 z = \frac{V_0}{2}.$$

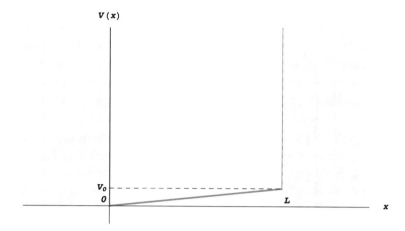

Fig. 6.2 Linear perturbation to a potential well

The change is the same for all eigenvalues. We will therefore have

$$E_n^0 + E_n^1 = \frac{\hbar^2 \pi^2 n^2}{2ma^2} + \frac{V_0}{2}, \qquad \forall n = 1, 2, \ldots$$

6.3 Particle on a Segment: Sinusoidal Perturbation

A particle of mass m is confined on a segment by an infinite potential well. Its bottom
is modified from

$$V(x) = 0 \qquad \text{for} \quad 0 < x < L$$

to

$$V'(x) = V_0 \sin(\frac{\pi x}{L}) \qquad \text{for} \quad 0 < x < L.$$

Calculate the first-order changes in the energy levels.

Solution

In the absence of the sinusoidal perturbation, the energy eigenvalues and eigenfunctions are

$$E_n = \frac{\hbar^2 \pi^2 n^2}{2mL^2}, \qquad \psi_n(x) = \sqrt{\frac{2}{L}} \sin \frac{n\pi x}{L}, \qquad n = 1, 2, 3, \ldots.$$

The first-order energy shifts are given by

$$E_n^1 = \int_0^L dx \, \psi_n^*(x) \, V'(x) \, \psi_n(x) =$$

$$= \frac{2V_0}{L} \int_0^L dx \, \sin^2 \frac{n\pi x}{L} \sin \frac{\pi x}{L} =$$

$$= \frac{2V_0}{\pi} \left[\frac{1}{2} \int_0^\pi d\alpha \, \sin \alpha - \frac{1}{2} \int_0^\pi d\alpha \, \cos 2n\alpha \, \sin \alpha \right] =$$

$$= \frac{2V_0}{\pi} \left\{ 1 - \frac{1}{2} \int_0^\pi d\alpha \, \frac{1}{2} \left[\sin(1 + 2n)\alpha + \sin(1 - 2n)\alpha \right] \right\} =$$

$$= \frac{2V_0}{\pi} \left[1 - \frac{1}{4} \left(\frac{2}{1 + 2n} + \frac{2}{1 - 2n} \right) \right] =$$

$$= \frac{8V_0 n^2}{\pi (4n^2 - 1)}.$$

6.4 Particle on a Segment in the Presence of a Dirac-δ Potential

A particle of mass m is in an infinite well potential of width a in one dimension:

$$V(x) = \begin{cases} 0, & \text{se } 0 \le x \le L, \\ \infty, & \text{elsewhere.} \end{cases}$$

It is subject to a perturbation due to the potential

$$W(x) = L\omega_0 \, \delta \left(x - \frac{L}{2} \right),$$

where ω_0 is a real constant.

(a) Calculate, up to the first order in ω_0, the changes in the energy levels produced by $W(x)$.

(b) Solve the problem exactly, showing that the energy values are obtained by one of the following two equations:

$$\sin(kL/2) = 0$$

or

$$\tan(kL/2) = -\frac{\hbar^2 k}{mL\omega_0}.$$

Discuss these results with regard to the sign and the size of ω_0.

(c) Show that, in the limit $\omega_0 \to 0$, the results of point a) are recovered.

Solution

(a) In the absence of perturbation, the energy eigenvalues and eigenfunctions are:

$$E_n = \frac{\hbar^2 \pi^2 n^2}{2mL^2}, \quad \psi_n(x) = \sqrt{\frac{2}{L}} \sin \frac{n\pi x}{L}, \quad n = 1, 2, 3, \ldots.$$

The first-order change in energy is given by

$$E_n^{(1)} = \int_0^L dx \, \psi_n^*(x) W(x) \psi_n(x) =$$

$$= \frac{2}{L} L\omega_0 \int_0^L dx \sin^2 \frac{n\pi x}{L} \, \delta(x - \frac{L}{2}) =$$

$$= 2\omega_0 \sin^2 \frac{n\pi}{2} = \begin{cases} 0 & \text{for even } n, \\ 2\omega_0 & \text{for odd } n. \end{cases}$$

(b) To find the exact solution, we impose that the wave function, excluding the point $x = L/2$, has the form of a null-potential solution vanishing in $x = 0$ and $x = L$. Setting $k = \sqrt{\frac{2mE}{\hbar^2}}$, we have

$$\psi(x) = \begin{cases} A \sin kx & \text{per } 0 \le x \le L/2, \\ B \sin k(L - x) & \text{per } L/2 \le x \le L. \end{cases}$$

Wave function continuity implies that

$$\psi(\tfrac{L}{2}^-) = \psi(\tfrac{L}{2}^+) \quad \Rightarrow \quad A \sin \tfrac{kL}{2} = B \sin \tfrac{kL}{2},$$

i.e.

$$A = B \quad \text{or} \quad A \ne B \text{ and } \sin \tfrac{kL}{2} = 0.$$

Instead, the first derivative must be discontinuous, due to the presence of the δ in the potential (see (2.26)):

$$\psi'\left(\frac{L}{2}^+\right) - \psi'\left(\frac{L}{2}^-\right) = \frac{2m}{\hbar^2} L\omega_0 \psi\left(\frac{L}{2}\right).$$

We distinguish two cases:

1. $A \ne B$ and $\sin \frac{kL}{2} = 0$. The latter relationship implies that

$$\frac{kL}{2} = n\pi \quad \Rightarrow \quad k = \frac{2n\pi}{L} \quad \text{con } n = 1, 2, \ldots$$

Thus, even in the exact calculation, we find the spectrum part of the infinite potential well corresponding to eigenfunctions that are odd with respect to

$x = L/2$. These eigenfunctions are not modified by the perturbation. Indeed, from the condition on ψ', we obtain

$$-Ak\cos\frac{kL}{2} - Bk\cos\frac{kL}{2} = 0 \quad \Rightarrow \quad A = -B.$$

The wave function in the range $L/2 \leq x \leq L$ is given by

$$\psi(x) = -A\sin k(L - x) = A\sin k(x - L) =$$
$$= A\sin kx\cos kL - A\cos kx\sin kL = A\sin kx,$$

because $\cos kL = 1$ and $\sin kL = 0$. We can say that, since these wave functions vanish in $x = L/2$, they do not feel the presence of the new potential term that, solely at this point, is not null.

2. $A = B$. In this case, the corresponding eigenfunctions are even with respect to $x = L/2$. From the condition on ψ', we obtain

$$-Ak\cos\frac{kL}{2} - Ak\cos\frac{kL}{2} = \frac{2m}{\hbar^2}L\omega_0 A\sin\frac{kL}{2},$$

that is,

$$\tan\frac{kL}{2} = -\frac{\alpha}{\omega_0}\frac{kL}{2}, \tag{6.1}$$

where

$$\alpha = \frac{2\hbar^2}{mL^2}.$$

To find this part of the spectrum, we solve this equation graphically, looking for the intersections between the two curves

$$y = \tan\frac{kL}{2} \quad \text{and} \quad y = -\frac{\alpha}{\omega_0}\frac{kL}{2}.$$

In Fig. 6.3, solutions for two opposite values of ω_0 are reported. For $\omega_0 \to 0$, these solutions tend to values (marked with a circle in the figure)

$$\frac{kL}{2} = (2j + 1)\frac{\pi}{2} \quad \Rightarrow \quad k = \frac{(2j + 1)\pi}{L}, \quad \text{con } j = 0, 1, \ldots,$$

that is,

$$E = \frac{\hbar^2\pi^2n^2}{2mL^2}, \quad n = 1, 3, 5, \ldots.$$

This returns the eigenvalues with the odd n of the unperturbed case. We note that the solution that, in the limit $\omega_0 \to 0^-$, corresponds to $k = \pi/L$ appears only if

$$\frac{\alpha}{|\omega_0|} > 1 \quad \Rightarrow \quad |\omega_0| < \frac{2\hbar^2}{mL^2}.$$

(c) As already seen, the solutions with n even are not modified by the presence of the δ potential. We now want to show that, in the case of an odd n, the perturbative solution corresponds to the exact solution's series expansion truncated to the first order in ω_0.

Consider, for example, the ground state with $n = 1$ and suppose that ω_0 is small and negative. The intersection of the two curves will be at $\frac{kL}{2}$, a little lower than $\frac{\pi}{2}$. Let us set, therefore,

$$\frac{kL}{2} = \frac{\pi}{2} - z, \tag{6.2}$$

with z small positive. Equation (6.1) becomes

$$\frac{\pi}{2} - z = \arctan\left[-\frac{\alpha}{\omega_0}\left(\frac{\pi}{2} - z\right)\right].$$

We expand the second member as a series in ω_0 and stop at the 1st order, obtaining

$$\frac{\pi}{2} - z = \frac{\pi}{2} + \frac{4\omega_0}{2\alpha(\pi - 2z)},$$

i.e.,

$$\alpha(\pi - 2z)z + 2\omega_0 = 0,$$

and, disregarding terms in z^2, we get the intersection point for ω_0 small:

$$z = -\frac{2\omega_0}{\alpha\pi}.$$

Substituting this value in (6.2), we find that

$$k = \frac{1}{L}\left(\pi + \frac{4\omega_0}{\alpha\pi}\right),$$

corresponding to an energy

$$E_1 = \frac{\hbar^2 k^2}{2m} = \frac{\hbar^2}{2mL^2}\left(\pi + \frac{4\omega_0}{\alpha\pi}\right)^2 \simeq \frac{\hbar^2\pi^2}{2mL^2} + 2\omega_0,$$

where we neglected the terms in ω_0^2 and replaced the value of α. This result is precisely the one obtained first-order Perturbation Theory for the ground state energy.

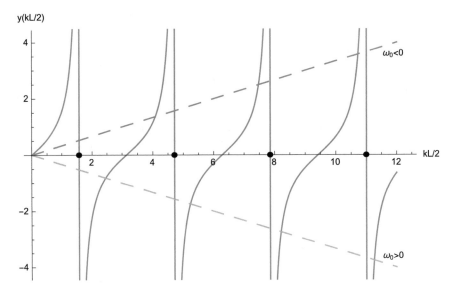

Fig. 6.3 δ perturbed potential well: graphic solution of the equation for the energy levels relative to even wave functions. The dashed lines correspond to two opposite values of ω_0

6.5   Particle in a Square: Coupling the Degrees of Freedom

Calculate the energy eigenfunctions and eigenvalues for a particle of mass m confined in a square of side L:
$$0 \le x \le L, \qquad 0 \le y \le L.$$

Then, introduce the perturbation $\mathcal{H}_1 = C\,xy$ and find the corrections up to the first order in the ground level and the first excited level.

Solution

The wave functions space is the tensor product of the spaces relative to two potential wells in normal directions. Unperturbed eigenvalues and eigenfunctions are given by:

$$\langle x, y | k, n \rangle = \psi_{k,n}(x, y) = \psi_k(x)\,\psi_n(y) = \sqrt{\tfrac{2}{L}} \sin \tfrac{k\pi x}{L} \sqrt{\tfrac{2}{L}} \sin \tfrac{n\pi y}{L},$$
$$E^0_{k,n} = E^0_k + E^0_n = \tfrac{\pi^2 \hbar^2 (k^2 + n^2)}{2mL^2},$$
$$\text{with } n, k = 1, 2, \ldots$$

To calculate the effects of the perturbation, we need the following matrix elements:

$$(\psi_1(x), x\,\psi_1(x)) = \frac{2}{L} \int_0^L dx\, x\, \sin^2 \frac{\pi x}{L} = \frac{L}{\pi^2} \int_0^\pi dx\, x\, (1 - \cos 2x) = \frac{L}{2},$$

$$(\psi_2(x), x\,\psi_2(x)) = \frac{2}{L} \int_0^L dx\, x\, \sin^2 \frac{2\pi x}{L} = \frac{2}{L} \frac{L^2}{\pi^2} \int_0^{2\pi} dx\, x\, \sin^2 x = \frac{L}{2},$$

$$(\psi_1(x), x\,\psi_2(x)) = \frac{2}{L} \int_0^L dx\, x\, \sin \frac{\pi x}{L} \sin \frac{2\pi x}{L} = \frac{2}{L} \frac{L^2}{\pi^2} \int_0^\pi dx\, x\, \sin x\, \sin 2x =$$

$$= \frac{4L}{\pi^2} \int_0^\pi dx\, x\, \sin^2 x\, \cos x = \frac{1}{3} \frac{4L}{\pi^2} \int_0^\pi d\cos x\, (1 - \cos^2 x) =$$

$$= -\frac{16}{9} \frac{L}{\pi^2}.$$

For the ground state, which is non-degenerate, we have

$$E^1_{1,1} = \langle 1, 1|\mathcal{H}_1|1, 1\rangle = (\psi_{1,1}(x, y), C\, xy\, \psi_{1,1}(x, y)) = C\left(\psi_1(x), x\,\psi_1(x)\right)^2 = \frac{CL^2}{4}.$$

Now consider the second level. We are in the presence of degeneracy between the $\psi_{1,2}$ and $\psi_{2,1}$ states. Therefore, we need to calculate the eigenvalues of the matrix

$$\begin{pmatrix} \langle 1, 2|\mathcal{H}_1|1, 2\rangle & \langle 1, 2|\mathcal{H}_1|2, 1\rangle \\ \langle 2, 1|\mathcal{H}_1|1, 2\rangle & \langle 2, 1|\mathcal{H}_1|2, 1\rangle \end{pmatrix} = \begin{pmatrix} A & B \\ B & A \end{pmatrix},$$

where

$$A = (\psi_{1,2}(x, y), C\, xy\, \psi_{1,2}(x, y)) = C\left(\psi_1(x), x\,\psi_1(x)\right)(\psi_2(y), y\,\psi_2(y)) = \frac{CL^2}{4},$$

$$B = (\psi_{1,2}(x, y), C\, xy\, \psi_{2,1}(x, y)) = C\left(\psi_1(x), x\,\psi_2(x)\right)^2 = \frac{256\,CL^2}{81\pi^4}.$$

The eigenvalues are $A \pm B$ and give the corrections to the first excited state:

$$E^1_{1,2} = CL^2\left(\frac{1}{4} \pm \frac{256}{81\pi^4}\right),$$

breaking up the degeneracy.

6.6 Particle on a Circumference in the Presence of Perturbation

An m mass particle is free to move around a circumference of radius R. The potential

$$V(\theta) = V_0 \sin\theta \cos\theta,$$

is applied, where θ is the angular coordinate on the circumference. Calculate the energy first-order corrections due to the presence of V. Identify the wave functions of the unperturbed system that diagonalize the matrix corresponding to V in each eigenspace and calculate the energy levels at the second perturbative order.

Solution

In the absence of perturbation, the system Hamiltonian contains only the kinetic term of rotation around the circumference center. By imposing the periodicity condition on the wave function, we find that

$$E_n = \frac{n^2 \hbar^2}{2m R^2}, \quad \psi_n(\theta) = \frac{1}{\sqrt{2\pi}} e^{in\theta}, \quad \text{con } n = 0, \pm 1, \dots$$

The ground level is not degenerate, while all the others are doubly degenerate. To calculate the required corrections, we first evaluate the perturbation matrix elements:

$$\begin{aligned}
\langle n|V|m \rangle &= \frac{V_0}{2\pi} \int_0^{2\pi} d\theta \, e^{i(m-n)\theta} \sin\theta \, \cos\theta = \\
&= \frac{V_0}{4\pi} \int_0^{2\pi} d\theta \, e^{i(m-n)\theta} \sin 2\theta = \\
&= \frac{V_0}{8\imath\pi} \left[\int_0^{2\pi} d\theta \, e^{i(m-n+2)\theta} - \int_0^{2\pi} d\theta \, e^{i(m-n-2)\theta} \right] = \\
&= \frac{V_0}{4\imath} [\delta_{m,n-2} - \delta_{m,n+2}].
\end{aligned} \tag{6.3}$$

First, we compute the first and second corrections in the energy of the ground state:

$$E_0^1 = \langle 0|V|0 \rangle = 0,$$

$$E_0^2 = \sum_{n \neq 0} \frac{|\langle n|V|0 \rangle|^2}{E_0 - E_n} = \frac{V_0^2}{16} \left[\frac{1}{E_0 - E_2} + \frac{1}{E_0 - E_{-2}} \right] = -\frac{V_0^2 m R^2}{16\hbar^2}.$$

Now let us move on to calculate the I order corrections in the degenerate eigenvalues. We need to diagonalize the matrix that represents V in the 2-dimensional subspace subtended by each eigenvalue. From the expression for $\langle n|V|m \rangle$, it is immediately understood that only in the case of $n = 1$ will we have a non-null matrix:

$$\begin{pmatrix} \langle +1|V|+1 \rangle & \langle +1|V|-1 \rangle \\ \langle -1|V|+1 \rangle & \langle -1|V|-1 \rangle \end{pmatrix} = \begin{pmatrix} 0 & \frac{V_0}{4\imath} \\ -\frac{V_0}{4\imath} & 0 \end{pmatrix}.$$

This matrix has eigenvalues

$$E_1^{1,+} = \frac{V_0}{4}, \quad E_1^{1,-} = -\frac{V_0}{4},$$

and corresponding eigenkets

$$|\psi_+\rangle = \frac{1}{\sqrt{2}} (|+1\rangle + \imath |-1\rangle), \quad |\psi_-\rangle = \frac{1}{\sqrt{2}} (|+1\rangle - \imath |-1\rangle).$$

These eigenvalues are the I order corrections in the $n = 1$ energy level. For all of the other levels, as stated above, the matrices and, therefore, the corrections are void.

Regarding the second-order corrections, we first consider those related to the first excited level, using, for $n = \pm 1$, the basis consisting of $|\psi_+\rangle$ and $|\psi_-\rangle$. Taking into account the fact that, for $n \neq 1$,

$$\langle \psi_+|V|n\rangle = \frac{1}{\sqrt{2}} (\langle +1|V|n\rangle - \imath \langle -1|V|n\rangle) = \frac{1}{\sqrt{2}} \frac{V_0}{4\imath} (\delta_{n,-1} - \delta_{n,3} - \imath \delta_{n,-3} + \imath \delta_{n,1}),$$

and that, obviously, in the new basis, it results that $\langle \psi_+|V|\psi_-\rangle = 0$, we obtain

$$E_1^{2,+} = \sum_{n \neq \pm 1} \frac{|\langle \psi_+|V|n\rangle|^2}{E_1 - E_n} = \frac{V_0^2}{32} \left[\frac{1}{E_1 - E_3} + \frac{1}{E_1 - E_{-3}} \right] =$$

$$= \frac{2m V_0^2 R^2}{16\hbar^2} \frac{1}{1 - 9} = -\frac{m V_0^2 R^2}{64\hbar^2}.$$

The same result is obtained for $E_1^{2,-}$, because the coefficients of the expansion of $|\psi_-\rangle$ and $|\psi_+\rangle$ in the old basis have the same modulus.

For $n \neq \pm 1$, we do not need to diagonalize the matrices in the degenerate eigenvalues subspaces, because the off-diagonal elements, due to (6.3), vanish. Finally, we find that

$$E_n^2 = \sum_{m \neq n} \frac{|\langle m|V|n\rangle|^2}{E_n - E_m} = \frac{V_0^2}{16} \left[\frac{1}{E_n - E_{n+2}} + \frac{1}{E_n - E_{n-2}} \right] =$$

$$= \frac{m V_0^2 R^2}{16\hbar^2} \frac{1}{n^2 - 1}.$$

6.7 Two Weakly Interacting Particles on a Circumference

Two particles of mass m are bound to stay on a circumference of radius R. Calculate the energy levels and eigenfunctions.

Then, suppose turning on an interaction between the particles by the potential

$$V = V_0 \cos(\phi_1 - \phi_2),$$

where ϕ_1 and ϕ_2 are the angular coordinates that identify the position of the two particles on the circumference.

(a) Introduce the variables $\alpha = \frac{\phi_1 + \phi_2}{2}$ and $\beta = \phi_1 - \phi_2$. Write the Schrödinger equation, show that it is separable in the new variables and determine the periodicity properties of the wave function in terms of these variables.
(b) Finally, perturbatively calculate the first-order changes in the energy eigenvalues.

Solution

(a) Since the Hamiltonian can be written in the form

$$\mathcal{H} = \mathcal{H}_1 + \mathcal{H}_2 \quad \text{where} \quad \mathcal{H}_i = -\frac{\hbar^2}{2mR^2} \frac{\partial^2}{\partial \phi_i^2},$$

the Schrödinger equation separates in two equations each one depending on a single variable. The eigenfunctions must satisfy the periodicity condition in these variables:

$$\psi(\phi_i + 2\pi) = \psi(\phi_i).$$

The Eigenfunctions and eigenvalues of $\mathcal{H}$ are therefore given by

$$\psi_{k,l}(\phi_1, \phi_2) = \psi_k(\phi_1)\psi_l(\phi_2), \qquad E_{k,l} = E_k + E_l,$$

where

$$\psi_n(\phi) = \frac{1}{\sqrt{2\pi}} e^{in\phi}, \qquad E_n = \frac{\hbar^2 n^2}{2mR^2}, \qquad n = 0, \pm 1, \ldots$$

In the presence of the potential $V = V_0 \cos(\phi_1 - \phi_2)$, the Schrödinger equation does not separate into the variables ϕ_1 and ϕ_2.
Introducing, instead, the suggested variables

$$\alpha = \frac{\phi_1 + \phi_2}{2} \quad \text{and} \quad \beta = \phi_1 - \phi_2,$$

the Hamiltonian separates into two terms dependent on a single variable:

$$\mathcal{H} = \mathcal{H}_{CM} + \mathcal{H}_r,$$

where

$$\mathcal{H}_{CM} = -\frac{\hbar^2}{2MR^2} \frac{\partial^2}{\partial \alpha^2} \quad \text{and} \quad \mathcal{H}_r = -\frac{\hbar^2}{2\mu R^2} \frac{\partial^2}{\partial \beta^2} + V(\beta).$$

$\mathcal{H}_{CM}$ corresponds to the free circular motion of the Center of mass with mass $M = 2m$ and angular position $\alpha \in [0, 2\pi]$, while $\mathcal{H}_r$ corresponds to the motion of the reduced mass $\mu = m/2$ in the presence of the potential $V(\beta)$ with $\beta \in [0, 2\pi]$. In the absence of $V(\beta)$, we have the eigenfunctions

$$\Psi_{k,l}(\alpha, \beta) = \Psi_k(\alpha)\Phi_l(\beta) \quad \text{relative to the eigenvalues} \quad E_{k,l} = \frac{\hbar^2 k^2}{2MR^2} + \frac{\hbar^2 l^2}{2\mu R^2}$$

with

$$\Psi_n(\phi) = \frac{1}{\sqrt{2\pi}} e^{in\alpha}, \qquad \Phi_n(\beta) = \frac{1}{\sqrt{2\pi}} e^{in\beta}, \qquad n = 0, \pm 1, \ldots$$

(b) The perturbation concerns only the Hamiltonian $\mathcal{H}_r$. We note that, since

$$\langle \Phi_k | V | \Phi_{-k} \rangle = \frac{V_0}{2\pi} \int_0^{2\pi} e^{-ik\beta} \cos \beta \, e^{-ik\beta} = 0 \qquad \forall k,$$

both the correction to the ground state (not degenerate) and the corrections to the excited states (doubly degenerate) are null; in this last case, in fact, the matrices to be diagonalized all have zero elements. We conclude that there is no energy change at first order.

6.8 Charged Rotator in an Electric Field

A plane rotator is a system consisting of two rigidly connected particles that rotates in a plane around an axis perpendicular to it and passing through the center of mass.

(a) Let m be the reduced mass of the two particles and a their distance. Determine the energy eigenvalues and eigenfunctions.
(b) Suppose that the particles are charged, so that the system has an electric dipole moment $\mathbf{d}$ and that it is immersed in a weak uniform electric field $\mathbf{E}$ directed in the rotation plane. Considering the interaction with the electric field as a perturbation, evaluate the first non-zero correction to the energy levels.

Solution

(a) Having called $I = ma^2$ the moment of inertia, the Hamiltonian is given by a pure kinetic term:

$$\mathcal{H} = \frac{L^2}{2I} = \frac{L_z^2}{2I} = -\frac{\hbar^2}{2I} \frac{\partial^2}{\partial \varphi^2},$$

where we assumed that the plane of motion is the plane xy and φ is the angular coordinate in that plane. The eigenvalues and eigenfunctions of $\mathcal{H}$ are

$$E_m = \frac{\hbar^2 m^2}{2I}, \qquad \psi_m(\varphi) = \frac{1}{\sqrt{2\pi}} e^{im\varphi}, \qquad m = 0, \pm 1, \pm 2, \ldots$$

The eigenvalues are doubly degenerate for every m, except for $m = 0$.

(b) The perturbation is given, assuming the electric field $\mathbf{E}$ to be directed along the x axis, by

$$\mathcal{H}' = -d\,E\,\cos\varphi.$$

First of all, we calculate the generic matrix element of $\cos\varphi$ in the $\mathcal{H}$ basis:

$$\langle n|\cos\varphi|m\rangle = \frac{1}{2\pi}\int_0^{2\pi} d\varphi\, e^{i(m-n)\varphi}\frac{e^{i\varphi}+e^{-i\varphi}}{2} =$$

$$= \frac{1}{4\pi}\left[\int_0^{2\pi} d\varphi\, e^{i(m-n+1)\varphi} + \int_0^{2\pi} d\varphi\, e^{i(m-n-1)\varphi}\right] =$$

$$= \frac{1}{2}[\delta_{m,n-1} + \delta_{m,n+1}].$$

Both in case $n^2 = 0$ (the absence of degeneracy) and in case $n^2 \neq 0$, the first-order corrections to energy are null, because all of the matrix elements are so; for example,

$$E^1_{n,-n} = -d\,E\,\langle n|\cos\varphi|-n\rangle = -d\,E\,[\delta_{-n,n-1} + \delta_{-n,n+1}] = 0 \quad \forall n = 0, \pm 1, \ldots$$

The second-order correction must therefore be calculated:

$$E^2_n = (dE)^2 \sum_{m \neq n} \frac{|\langle n|\cos\varphi|m\rangle|^2}{E_n - E_m} =$$

$$= \frac{(dE)^2}{4}\left[\frac{1}{E_n - E_{n-1}} + \frac{1}{E_n - E_{n+1}}\right] =$$

$$= \frac{(dE)^2}{4}\frac{2I}{\hbar^2}\left[\frac{1}{n^2 - (n-1)^2} + \frac{1}{n^2 - (n+1)^2}\right] =$$

$$= I\left(\frac{dE}{\hbar}\right)^2 \frac{1}{4n^2 - 1}.$$

We note that the second-order correction has been calculated without considering the degeneracy, since the perturbation is diagonal in the subspaces subtended by each eigenvalue (in the sum terms in which the denominator is canceled, which are those with $m = -n$, numerators are also null).

Finally, we note that, since E^2_n still depends on n^2, degeneracy is not eliminated.

6.9 Plane Rotator: Corrections Due to Weight Force

A point-like particle of mass m is constrained, in a vertical plane, to rotate around a horizontal axis by means of a negligible mass rod of length l. By treating the weight force in a perturbative manner, calculate the 2nd order corrections for the energy levels.

Solution

The solution is similar to that for problem (6.8). The unperturbed Hamiltonian contains only kinetic energy

$$\mathcal{H}_0 = \mathcal{T} = \frac{L_z^2}{2I} = -\frac{\hbar^2}{2I} \frac{\partial^2}{\partial \phi^2},$$

where $I = ml^2$ is the moment of inertia and ϕ is the angle of rotation around the z axis that we identify with the rotation axis. Placing $\phi = 0$ as the angle relative to the lowest position and setting the potential energy to be zero at $\phi = \pi/2$, the perturbation can be written as

$$\mathcal{H}_1(\phi) = -mgl \cos \phi.$$

$\mathcal{H}_0$ eigenfunctions are those of L_z:

$$\psi_n(\phi) = \frac{1}{\sqrt{2\pi}} e^{in\phi}, \qquad n = 0, \pm 1, \pm 2, \ldots,$$

and corresponds to the eigenvalues

$$E_n^0 = \frac{n^2 \hbar^2}{2I},$$

which are doubly degenerate for $n \neq 0$.

We calculate before the matrix elements

$$
\begin{aligned}
\langle n | \mathcal{H}_1 | k \rangle &= -\frac{mgl}{2\pi} \int_0^{2\pi} d\phi \, e^{-in\phi} \cos\phi \, e^{ik\phi} = \\
&= -\frac{mgl}{2\pi} \int_0^{2\pi} d\phi \, e^{i(k-n)\phi} \frac{e^{i\phi} + e^{-i\phi}}{2} = \\
&= -\frac{mgl}{2} (\delta_{k,n+1} + \delta_{k,n-1}).
\end{aligned}
$$

The first-order correction is null

$$E_n^1 = 0,$$

because the correction for the ground state, the only non-degenerate, is null, while, for the other states, the matrices to be diagonalized all have zero elements.

The 2nd order corrections can be calculated from the non-degenerate theory, because, due to the presence of the δ's in the matrix elements, the only contributions are from terms with non-zero denominators:

$$
\begin{aligned}
E_n^2 &= \sum_{m \neq n} \frac{|\langle m | \mathcal{H}^1 | n \rangle|^2}{E_n^0 - E_m^0} = \\
&= \left(\frac{mgl}{2} \right)^2 \left[\frac{1}{E_n^0 - E_{n+1}^0} + \frac{1}{E_n^0 - E_{n-1}^0} \right] = \\
&= \frac{m^3 g^2 l^4}{\hbar^2} \frac{1}{4n^2 - 1}.
\end{aligned}
$$

6.10 Harmonic Oscillator: Anharmonic Correction

The following interaction is added to the elastic potential:

$$
\mathcal{H}^1 = \lambda x^4.
$$

Calculate the first-order perturbative shift in the energy levels.

Solution

For this calculation, we need the expectation value of x^4 in a stationary state of the harmonic oscillator. This calculation is the object of problem 2.18, but, for the reader's convenience, we repeat it here. Using the completeness relation and results from problem 2.17, we obtain

$$
\langle x^4 \rangle_j = \langle j | x^4 | j \rangle = \sum_{k=0}^{\infty} \langle j | x^2 | k \rangle \langle k | x^2 | j \rangle = \sum_{k=0}^{\infty} |\langle j | x^2 | k \rangle|^2 = \sum_{k=0}^{\infty} \frac{\hbar^2}{4m^2 \omega^2} \times
$$

$$
\times \left[\sqrt{k(k-1)}\, \delta_{k,j+2} + \sqrt{(k+1)(k+2)}\, \delta_{k,j-2} + (2k+1)\delta_{j,k} \right]^2.
$$

Developing the square, products of δ's with different indices do not contribute. So, we get

$$
\begin{aligned}
\langle x^4 \rangle_j &= \frac{\hbar^2}{4m^2 \omega^2} \left[j(j-1) + (j+1)(j+2) + (2j+1)^2 \right] = \\
&= \frac{3\hbar^2}{4m^2 \omega^2} \left[2j^2 + 2j + 1 \right].
\end{aligned}
$$

Now we can easily obtain the desired correction:

$$E_n^1 = \langle n^0|\mathcal{H}^1|n^0\rangle = \lambda\langle n^0|x^4|n^0\rangle = \frac{3\hbar^2\lambda}{4m^2\omega^2}\left[2n^2 + 2n + 1\right].$$

We note that the I order approximation is justified only for the lower levels, since, although λ is small, the correction grows as n^2.

6.11 Harmonic Oscillator: Cubic Correction

Calculate, by Perturbation Theory, the energy eigenvalues up to order A^2 for an m mass particle that moves along the x-axis under the action of the potential

$$V(x) = \frac{1}{2}kx^2 + Ax^3.$$

Solution

Since the solution for the harmonic potential is known, we can write

$$\mathcal{H} = \mathcal{H}_0 + \mathcal{H}_1, \quad \text{where } \mathcal{H}_0 = \frac{p^2}{2m} + \frac{1}{2}kx^2, \text{ and } \mathcal{H}_1 = Ax^3$$

and consider $\mathcal{H}_1$ as a perturbation.

Using formulas (A.13, A.15), it is easily found that

$$\langle m|x^3|n\rangle = \left(\frac{\hbar}{2m\omega}\right)^{3/2}\langle m|(a + a^\dagger)^3|n\rangle =$$

$$= \left(\frac{\hbar}{2m\omega}\right)^{3/2}\left\{\sqrt{n(n-1)(n-2)}\,\delta_{m,n-3} + 3\sqrt{n^3}\,\delta_{m,n-1} + \right.$$

$$\left. +3\sqrt{(n+1)^3}\,\delta_{m,n+1} + \sqrt{(n+1)(n+2)(n+3)}\,\delta_{m,n+3}\right\}.$$

The first-order shift in the unperturbed nth energy level

$$E_n^0 = (n + \frac{1}{2})\hbar\omega$$

is given by (A.75):

$$E_n^1 = A\langle n|x^3|n\rangle = 0.$$

The second-order change is given by (A.76):

$$E_n^2 = A^2 \sum_{m \neq n} \frac{|\langle m|x^3|n\rangle|^2}{E_n^0 - E_m^0} = A^2 \left(\frac{\hbar}{2m\omega}\right)^3 \frac{1}{\hbar\omega} \left\{ \frac{n(n-1)(n-2)}{3} + \right.$$

$$\left. + \frac{(n+1)(n+2)(n+3)}{-3} + 9\frac{(n+1)^3}{-1} + 9n^3 \right\} =$$

$$= -\frac{\hbar^2 A^2}{8m^3\omega^4} (30n^2 + 30n + 11).$$

6.12   Harmonic Oscillator: Relativistic Correction

A particle of mass m moves in a harmonic potential

$$V(x) = \frac{1}{2}m\omega^2 x^2.$$

Its kinetic energy $T = \frac{p^2}{2m}$ can be considered as an approximation for small velocities with respect to the speed of light c of the relativistic expression:

$$T = \sqrt{p^2 c^2 + m^2 c^4}.$$

Determine, by Perturbation Theory, the correction in the energy levels at the order $\frac{1}{c^2}$.

Solution

The series expansion of the relativistic kinetic energy is

$$T = mc^2 \sqrt{1 + \frac{p^2}{m^2 c^2}} \simeq mc^2 + \frac{p^2}{2m} - \frac{p^4}{8m^3 c^2},$$

where we neglected terms of order above $(\frac{p}{c})^4$. The first term is the rest-energy and only redefines the arbitrary additive constant energy. In this approximation, the harmonic oscillator Hamiltonian is modified by a perturbative term

$$\mathcal{H}_1 = -\frac{p^4}{8m^3 c^2}.$$

To calculate the changes in the energy levels at order $\frac{1}{c^2}$, we use first-order Perturbation Theory in $\mathcal{H}_1$. We must therefore calculate the diagonal elements of the p^4 matrix in the basis of the harmonic oscillator energy. This could be done by using the expression for p in terms of creation and destruction operators or using the wave functions in the position representation. We will follow another path that brings us back to the calculation already made of the matrix elements of x^2 and x^4 (see problems 2.17 and 2.18). We can rewrite $\mathcal{H}_1$ in the form

$$\mathcal{H}_1 = -\frac{1}{2mc^2}\left(\frac{p^2}{2m}\right)^2 = -\frac{1}{2mc^2}\left(\mathcal{H}_0 - \frac{1}{2}m\omega^2 x^2\right)^2.$$

The required matrix elements are given by

$$\langle n|\mathcal{H}_1|n\rangle = -\frac{1}{2mc^2}\left[\langle n|\mathcal{H}_0^2|n\rangle - \frac{1}{2}m\omega^2\left(\langle n|\mathcal{H}_0 x^2|n\rangle - \langle n|x^2\mathcal{H}_0|n\rangle\right) + \frac{1}{4}m^2\omega^4\langle n|x^4|n\rangle\right] =$$

$$= -\frac{1}{2mc^2}\left(E_n^2 - m\omega^2 E_n\langle n|x^2|n\rangle + \frac{1}{4}m^2\omega^4\langle n|x^4|n\rangle\right) =$$

$$= -\frac{1}{2mc^2}\left[E_n^2 - m\omega^2 E_n\frac{E_n}{m\omega^2} + \frac{1}{4}m^2\omega^4\frac{3\hbar^2}{4m^2\omega^2}\left(2n^2 + 2n + 1\right)\right] =$$

$$= -\frac{3\hbar^2\omega^2}{32mc^2}\left(2n^2 + 2n + 1\right).$$

6.13 Anisotropic Harmonic Oscillator

A three-dimensional harmonic oscillator has an elastic constant k' along the z axis slightly different from the k constants along the x and y axes, i.e., its potential energy is

$$V(x, y, z) = \frac{1}{2}k(x^2 + y^2) + \frac{1}{2}k'z^2.$$

(a) Identify the ground state wavefunction. Note that it does not represent a definite angular momentum state. Why?
(b) At the first order in $(k - k')$, what are the angular momentum states other than 0 present in the ground state?

Solution

(a) The wave function space is the tensor product of the spaces relative to the oscillators arranged along the three axes. Placing

$$\omega = \sqrt{\frac{k}{m}} \quad \text{and} \quad \omega' = \sqrt{\frac{k'}{m}}$$

and, using (A.16)'s expressions for the harmonic oscillator eigenfunctions, we get

$$\psi_{0,0,0}(x,\,y,\,z) = \left(\frac{m\omega}{\pi\hbar}\right)^{\frac{1}{2}} \left(\frac{m\omega'}{\pi\hbar}\right)^{\frac{1}{4}} e^{-\frac{m\omega}{2\hbar}(x^2+y^2)}\, e^{-\frac{m\omega'}{2\hbar}z^2} =$$

$$= \left(\frac{m\omega}{\pi\hbar}\right)^{\frac{1}{2}} \left(\frac{m\omega'}{\pi\hbar}\right)^{\frac{1}{4}} e^{-\frac{m\omega}{2\hbar}(x^2+y^2+z^2)}\, e^{-\frac{m(\omega'-\omega)}{2\hbar}z^2} =$$

$$= \left(\frac{m\omega}{\pi\hbar}\right)^{\frac{1}{2}} \left(\frac{m\omega'}{\pi\hbar}\right)^{\frac{1}{4}} e^{-\frac{m\omega}{2\hbar}(r^2)}\, e^{-\frac{m(\omega'-\omega)}{2\hbar}r^2\cos^2\theta}.$$

The exponential dependence on $\cos^2\theta$ indicates the presence of contributions from all of the eigenfunctions of L^2. ψ does not depend on ϕ instead, so it is a L_z eigenfunction corresponding to the null eigenvalue.

(b) For the perturbative calculation, let's pose that

$$V(x,\,y,\,z) = V_0(x,\,y,\,z) + V_1(z),$$

where

$$V_0(x,\,y,\,z) = V(r) = \frac{1}{2}kr^2 \quad \text{and} \quad V_1(z) = \frac{1}{2}(k'-k)z^2.$$

At first order, we have, taking into account the factorization properties of the wave functions and problem 2.17, the following results:

$$E_0^1 = \langle 000|V_1|000\rangle = \frac{1}{2}(k'-k)\int dz\, z^2|\psi_0(z)|^2 = \frac{1}{2}(k'-k)\langle z^2\rangle_0 = \frac{\hbar}{4}\frac{k'-k}{\sqrt{km}}.$$

Since

$$\langle j|z^2|0\rangle = \frac{\hbar}{2m\omega}[\sqrt{2}\,\delta_{j,2} + \delta_{j,0}],$$

we get

$$|0^1\rangle = \sum_{m\neq 0}\frac{\langle m|V_1|0\rangle}{E_0 - E_m}|m\rangle = -\frac{1}{2}(k'-k)\left[\frac{\sqrt{2}\hbar}{2m\omega}\frac{1}{2\hbar\omega}|2\rangle\right] = -\frac{k'-k}{4\sqrt{2}\,k}|2\rangle$$

and, finally,

$$\psi_0'(x,\,y,\,z) = \psi_0(x)\psi_0(y)\left[\psi_0(z) - \frac{k'-k}{4\sqrt{2}\,k}\psi_2(z)\right] =$$

$$= \psi_0(x,\,y,\,z)\left[1 - \frac{1}{2\sqrt{2}}(4\xi^2 - 2)\frac{k'-k}{4\sqrt{2}\,k}\right],$$

where it was taken into account that, using (A.16, A.20),

$$\psi_2(z) = \psi_0(z)\left[\frac{1}{2\sqrt{2}}(4\xi^2 - 2)\right], \quad \text{where} \quad \xi = \sqrt{\frac{m\omega}{\hbar}}z.$$

Since $z^2 = r^2\cos^2\theta$, the wave function at I order takes contributions from the $\ell = 0, 2$ angular momentum states, as can be seen from (A.40).

6.14 Charged Harmonic Oscillator in an Electric Field

A particle of mass m and charge q subject to an elastic force is placed in a constant electric field, giving rise to a potential energy:

$$V(x) = \frac{1}{2}m\omega^2 x^2 - q\mathcal{E}x.$$

(a) Calculate, in Perturbation Theory, the changes at the first and second orders produced in the energy levels by the presence of the electric field and the changes at first order to the corresponding eigenket.
(b) Compare these perturbative results with the exact solution.

Solution

(a) Having introduced the following notation for the Hamiltonian,

$$\mathcal{H} = \mathcal{H}_0 + \mathcal{H}_1, \quad \text{where} \quad \mathcal{H}_0 = \frac{p^2}{2m} + \frac{1}{2}m\omega^2 x^2 \quad \text{and} \quad \mathcal{H}_1 = -q\mathcal{E}x,$$

we can immediately calculate the first-order corrections to energy levels using the result for $\langle x\rangle_n$ found in problem 2.16:

$$E_n^1 = 0, \quad \forall n \in N.$$

We now calculate the first-order correction for eigenkets, always using the results from problem 2.16:

$$|n\rangle = |n^0\rangle - q\mathcal{E}\sum_{m \neq n}\frac{\langle m^0|x|n^0\rangle}{E_n^0 - E_m^0}|m^0\rangle =$$

$$= |n^0\rangle - q\mathcal{E}\sqrt{\frac{\hbar}{2m\omega}}\left[\frac{\sqrt{n}\,|(n-1)^0\rangle}{\hbar\omega(n-n+1)} + \frac{\sqrt{n+1}\,|(n+1)^0\rangle}{\hbar\omega(n-n-1)}\right] =$$

$$= |n^0\rangle + q\mathcal{E}\sqrt{\frac{1}{2m\hbar\omega^3}}\left[\sqrt{n+1}\,|(n+1)^0\rangle - \sqrt{n}\,|(n-1)^0\rangle\right].$$

The effect of the perturbation at first order is therefore that of mixing each stationary state with the adjacent ones.

The second-order change in the energy levels is given by:

$$E_n^2 = \sum_{m \neq n} \frac{|\langle m^0|\mathcal{H}^1|n^0\rangle|^2}{E_n^0 - E_m^0} = q^2 \mathcal{E}^2 \frac{\hbar}{2m\omega}\left[\frac{n+1}{-\hbar\omega} + \frac{n}{\hbar\omega}\right] = -\frac{q^2\mathcal{E}^2}{2m\omega^2}.$$

(b) This problem can be solved exactly, enabling us to check Perturbation Theory results. Indeed,

$$\mathcal{H} = \frac{p^2}{2m} + \frac{1}{2}m\omega^2 x^2 - q\mathcal{E}x = \frac{p^2}{2m} + \frac{1}{2}m\omega^2\left(x - \frac{q\mathcal{E}}{m\omega^2}\right)^2 - \frac{1}{2}\frac{q^2\mathcal{E}^2}{m\omega^2}.$$

This Hamiltonian describes a harmonic oscillator whose rest position is translated into $x = \frac{q\mathcal{E}}{m\omega^2}$ and whose energy is shifted by a constant value $-\frac{1}{2}\frac{q^2\mathcal{E}^2}{m\omega^2}$. The translation of the center of oscillation does not affect the spectrum that derives only from the unchanged commutator's algebra:

$$[x - \frac{q\mathcal{E}}{m\omega^2}, p] = [x, p] = i\hbar.$$

Due to the potential energy shift, the energy spectrum is given by:

$$E_n = E_n^0 - \frac{q^2\mathcal{E}^2}{2m\omega^2}.$$

Therefore, the second-order perturbative contribution agrees with the exact result.

6.15   Harmonic Oscillator: Second Harmonic Potential I

The interaction

$$V(x) = \frac{1}{2}m\alpha^2 x^2$$

is added to the harmonic potential.

(a) Calculate the energy levels changes at first and second perturbative order.
(b) Compare the result with the exact value.

Solution

(a) Preliminarily, we calculate the matrix elements of the perturbation $V(x)$ in the base given by the unperturbed energy eigenstates. Using formulas (A.14, A.13), we easily obtain

$$\langle n|x^2|m \rangle = \frac{\hbar}{2m\omega} \left[\sqrt{n(n-1)}\, \delta_{n,m+2} + \sqrt{(n+1)(n+2)}\, \delta_{n,m-2} + (2n+1)\, \delta_{n,m} \right].$$

The first-order correction in the n-th level is given by

$$E_n^1 = \langle n|V|n \rangle = \frac{1}{2} m\alpha^2 \frac{\hbar}{2m\omega} (2n+1) = \frac{1}{2}\left(n+\frac{1}{2}\right)\frac{\hbar\alpha^2}{\omega}.$$

The second-order correction is, instead,

$$E_n^2 = \sum_{m \neq n} \frac{|\langle n|V|m \rangle|^2}{(n-m)\hbar\omega} =$$

$$= \left(\frac{\hbar\alpha^2}{4\omega}\right)^2 \left[\frac{n(n-1)}{2\hbar\omega} + \frac{(n+1)(n+2)}{-2\hbar\omega}\right] =$$

$$= -\frac{\hbar\alpha^4}{8\omega^3}\left(n+\frac{1}{2}\right).$$

(b) The exact result is obtained by considering an oscillator with frequency

$$\omega' = \sqrt{\omega^2 + \alpha^2}$$

whose energy levels are

$$E_n = (n+\frac{1}{2})\hbar\sqrt{\omega^2+\alpha^2} = (n+\frac{1}{2})\hbar\omega\sqrt{1+\frac{\alpha^2}{\omega^2}} =$$

$$= (n+\frac{1}{2})\hbar\omega\left[1 + \frac{1}{2}\frac{\alpha^2}{\omega^2} - \frac{1}{8}\left(\frac{\alpha^2}{\omega^2}\right)^2 + \ldots\right].$$

The terms of the perturbative series calculated earlier coincide with power expansion in the parameter α^2/ω^2 up to the second order.

6.16 Harmonic Oscillator: Second Harmonic Potential II

Solve the problem 6.15 in the wave function space.

Solution

The eigenfunctions of the unperturbed Hamiltonian are given by (A.16)

$$\phi_n(x) = c_n\, e^{-\frac{\xi^2}{2}} H_n(\xi) \quad \text{with} \quad c_n = \left(\frac{m\omega}{\pi\hbar}\right)^{\frac{1}{4}} \frac{1}{\sqrt{2^n n!}} \quad \text{and} \quad \xi = \sqrt{\frac{m\omega}{\hbar}}\, x,$$

where H_n is the n-th Hermite polynomial normalized by (A.18)

$$\int_{-\infty}^{+\infty} d\xi \, H_n^2(\xi) = \sqrt{\pi} \, 2^n \, n!. \tag{6.4}$$

To obtain the Perturbation Theory corrections, it is necessary to calculate the matrix elements of the potential, and, therefore, the matrix elements of x^2, in the unperturbed basis. They can be calculated using the recurrence relation of the Hermite polynomials (A.19):

$$\langle n|x^2|m\rangle = c_n c_m \int_{-\infty}^{+\infty} dx \, e^{-\xi^2} H_n(\xi) \, H_m(\xi) \, x^2 =$$

$$= \frac{1}{4} \, c_n c_m \left(\frac{m\omega}{\hbar}\right)^{-3/2} \int_{-\infty}^{+\infty} d\xi \, e^{-\xi^2} H_n(\xi) \, H_m(\xi) \, 4\xi^2 =$$

$$= \frac{1}{4} \, c_n c_m \left(\frac{m\omega}{\hbar}\right)^{-3/2} \int_{-\infty}^{+\infty} d\xi \, e^{-\xi^2} \Big[H_{n+1}(\xi) \, H_{m+1}(\xi) +$$

$$+ \, 4nm \, H_{n-1}(\xi) \, H_{m-1}(\xi) + 2m \, H_{n+1}(\xi) \, H_{m-1}(\xi) +$$

$$+ \, 2n \, H_{n-1}(\xi) \, H_{m+1}(\xi) \Big] =$$

$$= \frac{1}{4\sqrt{2^n n! 2^m m!}} \sqrt{\frac{m\omega}{\pi \hbar}} \left(\frac{m\omega}{\hbar}\right)^{-3/2} \Big[\sqrt{\pi} 2^{n+1} (n+1)! \delta_{n,m} +$$

$$+ \, 4n^2 \sqrt{\pi} 2^{n-1} (n-1)! \delta_{n,m} + 2m \sqrt{\pi} 2^{n+1} (n+1)! \delta_{n+2,m} +$$

$$+ \, 2n \sqrt{\pi} 2^{n-1} (n-1)! \delta_{n-2,m} \Big] =$$

$$= \frac{1}{2} \left(\frac{m\omega}{\hbar}\right)^{-1} \Big[(2n+1)\delta_{n,m} + \sqrt{n+1}(n+2)\delta_{n+2,m} + \sqrt{n(n-1)}\delta_{n-2,m} \Big].$$

The result found for the matrix elements is, obviously, the same one found through the technique of creation and destruction operators in problem 6.15; as a consequence, the perturbative corrections are the same.

6.17 Plane Harmonic Oscillator: Linear and Quadratic Correction

A plane harmonic oscillator has Hamiltonian

$$\mathcal{H}_0 = \frac{1}{2m} \, (p_x^2 + p_y^2) + \frac{1}{2} m\omega^2 \, (x^2 + y^2).$$

(a) Calculate the energy levels and their degeneration.

(b) Adding a perturbation $V_1 = \epsilon x$ to H_0, perturbatively compute corrections to the levels at the first and second orders.

(c) Adding a perturbation $V_2 = \epsilon x^2$ to H_0, perturbatively compute corrections to the levels at the first and second orders.

(d) Compare the results obtained in b) and c) with the respective exact results.

Solution

(a) The Hilbert space of the system is the tensor product of the spaces related to two equal frequency oscillators arranged along the x and y axes.
So, the energy spectrum is given by

$$E_n = \hbar\omega(n+1),$$

where n is the sum of two integers n_x and n_y. The level degeneracy is $n+1$.

(b) A perturbation exists only in the x direction:

$$E_{n_x}^1 = \epsilon \langle n_x | x | n_x \rangle = 0,$$

$$E_{n_x}^2 = -\frac{\epsilon^2}{2m\omega^2},$$

where we used the results from problem 6.14. Ultimately, the second-order change is the same for all levels:

$$E_n^2 = \hbar\omega(n+1) - \frac{\epsilon^2}{2m\omega^2}.$$

(c) Using the results from problem 6.15, we find that

$$E_{n_x}^1 = \frac{\hbar\epsilon}{m\omega}\left(n_x + \frac{1}{2}\right),$$

$$E_{n_x}^2 = -\frac{\hbar\epsilon^2}{2m^2\omega^3}\left(n_x + \frac{1}{2}\right).$$

Therefore, the second-order series expansion of the energy levels is

$$E_{n_x,n_y} = \hbar\omega(n_x + n_y + 1) + \frac{\hbar\epsilon}{m\omega}\left(n_x + \frac{1}{2}\right) - \frac{\hbar\epsilon^2}{2m^2\omega^3}\left(n_x + \frac{1}{2}\right) =$$

$$= \hbar\omega\left(\frac{1}{2} + \frac{\epsilon}{m\omega^2} - \frac{\epsilon^2}{2m^2\omega^4}\right)\left(n_x + \frac{1}{2}\right) + \hbar\omega\left(n_y + \frac{1}{2}\right).$$

There is no degeneracy, except for special values of ϵ.

(d) In the case of the perturbation V_1, we know (see problem 6.14) that the second-order perturbative correction coincides with the exact result. In the case of the perturbation V_2, we can write

$$\mathcal{H} = \frac{1}{2m}(p_x^2 + p_y^2) + \frac{1}{2}m\omega^2(x^2 + y^2) + \epsilon x^2 =$$

$$= \frac{1}{2m}p_x^2 + \frac{1}{2}m\omega'^2 x^2 + \frac{1}{2m}p_y^2 + \frac{1}{2}m\omega^2 y^2,$$

where

$$\omega'^2 = \omega^2 + \frac{2\epsilon}{m}.$$

Therefore, we have

$$E_{n_x,n_y} = \hbar\omega'(n_x + \frac{1}{2}) + \hbar\omega(n_y + \frac{1}{2}) =$$

$$= \hbar\omega\sqrt{1 + \frac{2\epsilon}{m\omega^2}}(n_x + \frac{1}{2}) + \hbar\omega(n_y + \frac{1}{2}) =$$

$$= \hbar\omega\left[1 + \frac{\epsilon}{m\omega^2} - \frac{1}{2}\left(\frac{\epsilon}{m\omega^2}\right)^2 + O(\epsilon^2)\right](n_x + \frac{1}{2}) + \hbar\omega(n_y + \frac{1}{2}).$$

The second-order series expansion coincides with the second-order ϵ expansion of the exact result.

6.18   Coupled Harmonic Oscillators

Consider two identical unidimensional harmonic oscillators of mass m and elastic constant k. They interact via a potential energy

$$\mathcal{H}_1 = \alpha x_1 x_2,$$

where x_1 and x_2 are the positions of the two oscillators.

(a) Determine the energy eigenvalues and eigenstates. (Hint: separate the motion of the center of mass from that of the reduced mass.)
(b) In the hypothesis in which $\alpha \ll k$, calculate the energy levels at the lowest perturbative order.
(c) Compare the two results.

Solution

(a) The system Hamiltoniano is

$$\mathcal{H} = \frac{p_1^2}{2m} + \frac{p_2^2}{2m} + \frac{1}{2}k x_1^2 + \frac{1}{2}k x_2^2 + \alpha x_1 x_2.$$

Introducing, as suggested, the variables

$$X = \frac{x_1 + x_2}{2} \quad \text{and} \quad x = x_1 - x_2,$$

the Hamiltonian becomes:

$$\mathcal{H} = \mathcal{H}_{CM} + \mathcal{H}_r$$

where

$$\mathcal{H}_{CM} = \frac{P^2}{2M} + (k + \alpha)X^2 \quad \text{and} \quad \mathcal{H}_r = \frac{p^2}{2\mu} + \frac{1}{4}(k - \alpha)x^2.$$

Here, P and p are the conjugate momenta of the new variables, $M = 2m$ is the total mass and $\mu = m/2$ is the reduced mass.

The system then presents a Hamiltonian sum of two terms related to two oscillating motions, one of elastic constant $2(k + \alpha)$ for the center of mass and one of constant constant $(k - \alpha)/2$ for the relative motion. The Schrödinger equation is therefore separable into the new variables. We note that the condition $\alpha \ll k$ should be verified, otherwise the relative motion would have potential energy not bounded from below.

The energy spectrum is therefore given by

$$E_{n_T, n_r} = E_{n_T} + E_{n_r} = \hbar\omega_T \left(n_T + \frac{1}{2} \right) + \hbar\omega_r \left(n_r + \frac{1}{2} \right),$$

where, said ω being the two oscillators' frequency in the absence of interaction, we set

$$\omega_T = \sqrt{\frac{2(k + \alpha)}{M}} = \sqrt{\frac{k + \alpha}{m}} = \sqrt{\frac{k}{m}} \sqrt{1 + \frac{\alpha}{k}} = \omega \sqrt{1 + \frac{\alpha}{m\omega^2}},$$

$$\omega_r = \sqrt{\frac{k - \alpha}{2\mu}} = \sqrt{\frac{k - \alpha}{m}} = \sqrt{\frac{k}{m}} \sqrt{1 - \frac{\alpha}{k}} = \omega \sqrt{1 - \frac{\alpha}{m\omega^2}}.$$

(b) Perturbative corrections can be calculated using both the old x_1 and x_2 coordinates and the new X and x coordinates. Here, we use the latter.

In the absence of interaction, the two oscillators concerning the center of mass and the relative coordinate, have the same frequency ω. Therefore, the energy levels are

$$E_n = \hbar\omega(n + 1), \quad \text{where} \quad n = n_T + n_r = 0, 1, \ldots$$

Each n level has a degeneracy of $n + 1$.

Using the new coordinates, the interaction potential, which constitutes the perturbation, becomes

$$\mathcal{H}_1 = \alpha\, x_1 x_2 = \alpha\, \frac{2X + x}{2}\, \frac{2X - x}{2} = \alpha\left(X^2 - \frac{x^2}{4}\right).$$

To calculate the first-order shift in the nth energy level, we need the eigenvalues of the perturbation matrix in the subspace of the ket $|n\rangle = |j, n - j\rangle = |j\rangle_X \otimes |n - j\rangle_x$, where the j index ranges from 0 to n. Using the results from problem 2.17, we get

$$(\mathcal{H}_1)_{j,k} = \alpha\langle j, n - j | x_1 x_2 | k, n - k\rangle =$$

$$= \alpha\langle j, n - j | (X^2 - \frac{x^2}{4}) | k, n - k\rangle =$$

$$= \alpha\left[\langle j | X^2 | k\rangle \delta_{j,k} - \frac{1}{4}\langle n - j | x^2 | n - k\rangle \delta_{j,k}\right] =$$

$$= \alpha\left[\langle k | X^2 | k\rangle \delta_{j,k} - \frac{1}{4}\langle n - k | x^2 | n - k\rangle \delta_{j,k}\right] =$$

$$= \alpha\left[\frac{\hbar}{2M\omega}(2k + 1) - \frac{1}{4}\frac{\hbar}{2\mu\omega}(2(n - k) + 1)\right]\delta_{j,k} =$$

$$= \frac{\alpha\hbar}{2m\omega}(2k - n)\delta_{j,k}.$$

The matrix is then diagonal and the $n + 1$ eigenvalues are the product of $\frac{\alpha\hbar}{2m\omega}$ times

$$-n, 2 - n, 4 - n, \ldots, n - 4, n - 2, n.$$

(c) Expanding the expressions for ω_T and ω_r in the power series of $\alpha/k = \alpha/m\omega^2$ and neglecting orders above the second one, we obtain

$$\omega_T \simeq \omega\left[1 + \frac{1}{2}\frac{\alpha}{m\omega^2} - \frac{1}{8}\left(\frac{\alpha}{m\omega^2}\right)^2\right],$$

$$\omega_r \simeq \omega\left[1 - \frac{1}{2}\frac{\alpha}{m\omega^2} - \frac{1}{8}\left(\frac{\alpha}{m\omega^2}\right)^2\right].$$

We can therefore approximate the total energy spectrum with the following expression:

$$E_{N_T, n_r} \simeq \hbar\omega(n_T + n_r + 1) + \frac{1}{2}\frac{\alpha\hbar}{m\omega}(n_T - n_r) - \frac{\hbar\omega}{8}\left(\frac{\alpha}{m\omega^2}\right)^2 (n_T + n_r + 1),$$

and, setting $n_T + n_r = n$,

$$E_{n,n_T} = \hbar\omega(n+1) + \frac{1}{2}\frac{\alpha\hbar}{m\omega}(2n_T - n) + O(\alpha^2), \quad \text{with } n_T = 0, 1, \ldots n.$$

The first-order term coincides with the first-order perturbative result, because the difference $2n_T - n$ assumes the proper values $-n, 2 - n, 4 - n, \ldots, n - 4, n - 2, n$.

6.19 Plane Harmonic Oscillator: Coupling the Degrees of Freedom

A particle of mass m moves in the plane xy subject to a harmonic potential with frequency ω:

$$\mathcal{H}_0 = \frac{1}{2m}(p_x^2 + p_y^2) + \frac{1}{2}m\omega^2(x^2 + y^2).$$

Introduce the perturbation

$$\mathcal{H}_1(x, y) = 2\lambda xy.$$

(a) Find eigenvalues and eigenstates of $\mathcal{H}_0$.
(b) Calculate first and second-order corrections in the energy of the ground state due to $\mathcal{H}_1$.
(c) Calculate first-order correction in the energy of the first and second excited level due to $\mathcal{H}_1$.

Solution

The problem is solved exactly in 6.18.

(a) With an obvious assignment of symbols, we can write

$$\mathcal{H}_0 = \mathcal{H}_0^x + \mathcal{H}_0^y.$$

The unperturbed Hamiltonian eigenvalues are

$$E_n^0 = \hbar\omega(n+1), \quad \text{where} \quad n = n_x + n_y \quad \text{and} \quad n_{x,y} = 0, 1, 2, \ldots$$

and their eigenstates, once those of the one-dimensional harmonic oscillator for each direction have been denoted by $|n_x\rangle$ and $|n_y\rangle$, are

$$|n^0\rangle = |n_x\rangle|n_y\rangle.$$

(b) The ground state is
$$|0^0\rangle = |0_x\rangle|0_y\rangle.$$

The first-order correction is

$$E_0^1 = \langle 0^0|2\lambda xy|0^0\rangle = 2\lambda\langle 0|x|0\rangle^2 = 2\lambda\left(\sqrt{\frac{\hbar}{2m\omega}}\langle 0|1\rangle\right)^2 = 0,$$

where we used (A.14, A.15).
At the second order we have, taking into account (A.76),

$$E_0^2 = 4\lambda^2 \sum_{m\neq 0} \frac{|\langle 0^0|xy|m^0\rangle|^2}{(0-m)\hbar\omega} =$$

$$= 4\lambda^2 \sum_{(n_x,n_y)\neq(0,0)} \frac{|\langle 0|x|n_x\rangle\langle 0|y|n_x\rangle|^2}{(-n_x-n_y)\hbar\omega} =$$

$$= \frac{4\lambda^2}{\hbar\omega}\left(\frac{\hbar}{2m\omega}\right)^2\frac{1}{-2} = -\frac{\lambda^2\hbar}{2m^2\omega^3}.$$

(c) The first and second excited levels are degenerate; therefore, the corrections are given by the eigenvalues of the matrix representing the perturbation in the subspace relative to the level. For simplicity's sake, we again indicate this matrix with the $\mathcal{H}_1$ symbol.
The first excited level is doubly degenerate. We have

$$\mathcal{H}_1 = \begin{pmatrix} \langle 10|\mathcal{H}_1|10\rangle & \langle 10|\mathcal{H}_1|01\rangle \\ \langle 01|\mathcal{H}_1|10\rangle & \langle 01|\mathcal{H}_1|01\rangle \end{pmatrix} = 2\lambda\begin{pmatrix} 0 & \frac{\hbar}{2m\omega} \\ \frac{\hbar}{2m\omega} & 0 \end{pmatrix},$$

eigenvalues of which (the corrections) are $\pm\frac{\lambda\hbar}{m\omega}$.
The second excited level is three times degenerate, and its eigenstates are

$$|20\rangle, \quad |11\rangle \quad |02\rangle \, ;$$

then, the matrix to diagonalize is

$$\mathcal{H}_1 = \begin{pmatrix} \langle 20|\mathcal{H}_1|20\rangle & \langle 20|\mathcal{H}_1|11\rangle & \langle 20|\mathcal{H}_1|02\rangle \\ \langle 11|\mathcal{H}_1|20\rangle & \langle 11|\mathcal{H}_1|11\rangle & \langle 11|\mathcal{H}_1|02\rangle \\ \langle 02|\mathcal{H}_1|20\rangle & \langle 02|\mathcal{H}_1|11\rangle & \langle 02|\mathcal{H}_1|02\rangle \end{pmatrix} = \frac{\lambda\hbar}{m\omega}\begin{pmatrix} 0 & \sqrt{2} & 0 \\ \sqrt{2} & 0 & \sqrt{2} \\ 0 & \sqrt{2} & 0 \end{pmatrix}.$$

Its eigenvalues are 0 and $\pm\frac{2\lambda\hbar}{m\omega}$.

6.20   Pendulum: Anharmonic Correction to Small Oscillations

Consider a pendulum of mass m and length l that oscillates in a vertical plane.

(a) Find the energy levels in the approximation of small oscillations.
(b) Consider the next approximation to that of the small oscillations and find the first-order perturbative corrections to the energy levels.

Solution

Having established that θ is the deviation angle with respect to the vertical, the pendulum Hamiltonian is given by

$$\mathcal{H} = \frac{1}{2} ml^2 \dot{\theta}^2 + mgl(1 - \cos\theta).$$

By developing the potential energy in power series around $\theta = 0$, we have

$$V(\theta) = mgl \left(\frac{\theta^2}{2} - \frac{\theta^4}{24} + O\left(\theta^6\right) \right).$$

(a) Within the limit of small oscillations, the Hamiltonian is approximated by that of a harmonic oscillator:

$$\mathcal{H}_0 = \frac{1}{2} ml^2 \dot{\theta}^2 + \frac{1}{2} mgl\theta^2 = \frac{p^2}{2m} + \frac{1}{2} m\omega^2 q^2,$$

where we introduced the displacement $q = l\theta$ with respect to the rest position and the frequency $\omega = \sqrt{\frac{g}{l}}$. The energy spectrum is therefore given by

$$E_n = (n + \frac{1}{2})\hbar\omega.$$

(b) Considering also the second order in θ^2, the potential energy is

$$V = mgl \left(\frac{\theta^2}{2} - \frac{\theta^4}{24} \right) = \frac{1}{2} m\omega^2 q^2 - \frac{1}{24} \frac{mg}{l^3} q^4.$$

In order to find the first-order perturbative corrections to the energy levels, remember that (2.18)

$$< q^4 >_j = \frac{3\hbar^2}{4m^2\omega^2} \left(2j^2 + 2j + 1\right).$$

Therefore, the required corrections are given by

$$E_n^1 = \langle n|\mathcal{H}_1|n\rangle = -\frac{1}{32}\frac{g\hbar^2}{ml^3\omega^2}\left(2n^2 + 2n + 1\right).$$

6.21 Degeneracy Breakdown in a Two-State System

The Hamiltonian for a two-state system has the form

$$\mathcal{H} = \mathcal{H}_0 + \lambda\mathcal{H}_1 = \begin{pmatrix} a & \lambda\Delta \\ \lambda\Delta & b \end{pmatrix}, \qquad (\lambda > 0).$$

(a) Solve the energy eigenvalues problem by exactly determining eigenvectors and eigenvalues.
(b) Assuming $\lambda|\Delta| \ll |a - b|$, solve the same problem in Perturbation Theory up to the first order in the eigenvectors and up to the second order in the eigenvalues.
(c) Assuming that, in the absence of perturbation, energy levels are almost degenerate,

$$|a - b| \ll \lambda|\Delta|$$

show that the results obtained by applying the first-order Degenerate Perturbation Theory $(a = b)$ are compatible with those derived from the exact calculation.

Solution

We immediately see that the eigenvectors and eigenvalues of

$$\mathcal{H}_0 = \begin{pmatrix} a & 0 \\ 0 & b \end{pmatrix}$$

are given by

$$|1^0\rangle = \begin{pmatrix} 1 \\ 0 \end{pmatrix} \text{ relative to } E_1^0 = a,$$

$$|2^0\rangle = \begin{pmatrix} 0 \\ 1 \end{pmatrix} \text{ relative to } E_2^0 = b.$$

(a) The eigenvalues of $\mathcal{H}$ are solutions to the characteristic equation

$$(a - \omega)(b - \omega) - \lambda^2\Delta^2 = 0,$$

$$\omega_\pm = \frac{a+b}{2} \pm \sqrt{\frac{(a-b)^2}{4} + \lambda^2\Delta^2}.$$

The relative eigenvectors are obtained solving the homogeneous systems

$$\begin{pmatrix} a & \lambda\Delta \\ \lambda\Delta & b \end{pmatrix} \begin{pmatrix} \alpha \\ \beta \end{pmatrix} = \omega_{\pm} \begin{pmatrix} \alpha \\ \beta \end{pmatrix}.$$

Thus, we obtain the equations

$$a\alpha + \lambda\Delta\beta = \omega_{\pm}\alpha,$$

$$\beta = \frac{\omega_{\pm}-a}{\lambda\Delta}\alpha,$$

$$|\psi_{\pm}\rangle = \frac{1}{\sqrt{\lambda^2\Delta^2+(\omega_{\pm}-a)^2}} \begin{pmatrix} \lambda\Delta \\ \omega_{\pm} - a \end{pmatrix},$$

where normalization has been imposed.

(b) Now we apply Perturbation Theory. Let us calculate the first-order corrections to energy eigenvalues:

$$E_1^1 = \lambda\langle 1^0|H_1|1^0\rangle = \lambda\Delta \begin{pmatrix} 1 & 0 \end{pmatrix} \begin{pmatrix} 0 & 1 \\ 1 & 0 \end{pmatrix} \begin{pmatrix} 1 \\ 0 \end{pmatrix} = 0,$$

$$E_2^1 = \lambda\langle 2^0|H_1|2^0\rangle = \lambda\Delta \begin{pmatrix} 0 & 1 \end{pmatrix} \begin{pmatrix} 0 & 1 \\ 1 & 0 \end{pmatrix} \begin{pmatrix} 0 \\ 1 \end{pmatrix} = 0.$$

The first-order corrections to energy eigenstates are

$$|1^1\rangle = |1^0\rangle + \lambda\frac{\langle 2^0|\mathcal{H}_1|1^0\rangle}{E_1^0-E_2^0}|2^0\rangle = |1^0\rangle + \lambda\frac{\Delta}{a-b}|2^0\rangle,$$

$$|2^1\rangle = |2^0\rangle + \lambda\frac{\langle 1^0|\mathcal{H}_1|2^0\rangle}{E_2^0-E_1^0}|1^0\rangle = |2^0\rangle - \lambda\frac{\Delta}{a-b}|1^0\rangle.$$

Ultimately, we calculate the second-order energy shifts:

$$E_1^2 = \frac{|\langle 1^0|\lambda\mathcal{H}_1|2^0\rangle|^2}{E_1^0 - E_2^0} = \lambda^2\frac{\Delta^2}{a - b},$$

$$E_2^2 = \frac{|\langle 2^0|\lambda\mathcal{H}_1|1^0\rangle|^2}{E_2^0 - E_1^0} = -\lambda^2\frac{\Delta^2}{a - b}.$$

These results coincide, of course, with the second-order term of the series expansion in λ of the exact eigenvalues. Indeed,

$$\omega_{\pm} = \frac{a+b}{2} \pm \frac{a-b}{2}\sqrt{1 + \frac{4\lambda^2\Delta^2}{(a-b)^2}} = \frac{a+b}{2} \pm \frac{a-b}{2}\left(1 + \frac{2\lambda^2\Delta^2}{(a-b)^2}\right) =$$

$$= \begin{cases} a + \frac{\lambda^2\Delta^2}{a-b}, \\ b - \frac{\lambda^2\Delta^2}{a-b}. \end{cases}$$

(c) In order to apply Degenerate Perturbation Theory in the case $(a = b)$, we need
to diagonalize the perturbation matrix:

$$det \begin{pmatrix} -E^1 & \lambda\Delta \\ \lambda\Delta & -E^{(1)} \end{pmatrix} = 0 \quad \Rightarrow \quad E^1 = \pm\lambda\Delta.$$

The energy eigenvalues up to first order are

$$E = a \pm \lambda\Delta.$$

To compare this result with the exact one, we note that, if

$$|a - b| \ll \lambda|\Delta|,$$

the following series expansion makes sense:

$$\omega_\pm = \frac{a+b}{2} \pm \lambda\Delta \sqrt{1 + \frac{(a-b)^2}{4\lambda^2\Delta^2}} = \frac{a+b}{2} \pm \lambda\Delta \pm \frac{1}{8}\frac{(a-b)^2}{\lambda^2\Delta^2} + \ldots =$$
$$= a \pm \lambda\Delta + O\left(\frac{(a-b)^2}{\lambda^2\Delta^2}\right).$$

This result coincides with the previous result, unless second-order corrections
in λ.

6.22   Fermion in a Magnetic Field

Consider the Hamiltonian of a spin 1/2 particle immersed in a uniform and constant
magnetic field **B**, which is obtained by ignoring the kinetic term:

$$\mathcal{H} = -\mu\hbar\mathbf{B} \cdot \boldsymbol{\sigma}.$$

Consider the case in which **B** lies on the xz plane with

$$\epsilon = \frac{B_x}{B_z} \ll 1.$$

(a) Determine, by Perturbation Theory, the eigenvalues of $\mathcal{H}$ up to the order ϵ^2
included, and the eigenstates up to the order ϵ.
(b) Determine the exact eigenvalues and eigenstates, comparing them with the prece-
dent results.

Solution

(a) We can write

$$\mathbf{B} = (\varepsilon B_z, 0, B_z)$$

and

$$\mathcal{H} = \mathcal{H}_0 + \mathcal{H}_1, \quad \text{where} \quad \mathcal{H}_0 = -\mu \hbar B_z \sigma_z \quad \text{and} \quad \mathcal{H}_1 = -\varepsilon \mu \hbar B_z \sigma_x.$$

The unperturbed Hamiltonian $\mathcal{H}_0$ is proportional to σ_z and therefore its eigenvalues are

$$E_-^0 = -\mu \hbar B_z \quad \text{relative to the eigenstate} \quad |-\rangle = |\sigma_z = +1\rangle = \begin{pmatrix} 1 \\ 0 \end{pmatrix},$$

$$E_+^0 = +\mu \hbar B_z \quad \text{relative to the eigenstate} \quad |+\rangle = |\sigma_z = -1\rangle = \begin{pmatrix} 0 \\ 1 \end{pmatrix}.$$

In the $\mathcal{H}_0$ representation, $\mathcal{H}_1$ is given by the matrix

$$\mathcal{H}_1 = \begin{pmatrix} \langle -|\mathcal{H}_1|-\rangle & \langle -|\mathcal{H}_1|+\rangle \\ \langle +|\mathcal{H}_1|-\rangle & \langle +|\mathcal{H}_1|+\rangle \end{pmatrix} = \begin{pmatrix} 0 & -\varepsilon \mu \hbar B_z \\ -\varepsilon \mu \hbar B_z & 0 \end{pmatrix}.$$

We immediately see that the first-order corrections in the eigenvalues are void. The first-order corrections in the eigenstates are given by

$$|n^1\rangle = \sum_{m \neq n} \frac{\langle m^0|\mathcal{H}_1|n^0\rangle}{E_n^0 - E_m^0} |m^0\rangle,$$

i.e.,

$$|E_-^1\rangle = -\frac{1}{2\mu \hbar B_z}(-\mu \hbar B_z \varepsilon)|E_+\rangle = +\frac{\varepsilon}{2} \begin{pmatrix} 0 \\ 1 \end{pmatrix},$$

$$|E_+^1\rangle = +\frac{1}{2\mu \hbar B_z}(-\mu \hbar B_z \varepsilon)|E_-\rangle = -\frac{\varepsilon}{2} \begin{pmatrix} 1 \\ 0 \end{pmatrix}.$$

The second-order corrections in the eigenstates are given by

$$E_n^2 = \sum_{m \neq n} \frac{|\langle m^0|\mathcal{H}_1|n^0\rangle|^2}{E_n^0 - E_m^0},$$

i.e.,

$$E_-^2 = -\frac{\mu \hbar B_z \varepsilon^2}{2}, \qquad E_+^2 = +\frac{\mu \hbar B_z \varepsilon^2}{2}.$$

(c) We now solve the exact eigenvalue problem for

$$\mathcal{H} = -\mu\hbar B_z \, \Lambda, \quad \text{where} \quad \Lambda = \begin{pmatrix} 1 & \varepsilon \\ \varepsilon & -1 \end{pmatrix}.$$

Λ eigenvalues are $\lambda_{1,2} = \pm\sqrt{1 + \varepsilon^2}$, so those of $\mathcal{H}$ are

$$E_{1,2} = \mp\mu\hbar B_z\sqrt{1 + \varepsilon^2},$$

which, if expanded in the power series of ε, coincide, up to the second order with the results found in Perturbation Theory.
The $\mathcal{H}$ eigenvectors are the same as those of Λ, i.e.,

$$|E_{1,2}\rangle = \frac{1}{(1 + \lambda_{1,2})^2 + \varepsilon^2}\begin{pmatrix} 1 + \lambda_{1,2} \\ \varepsilon \end{pmatrix}.$$

6.23 β Decay in a Hydrogenlike Atom

Find, in first-order Perturbation Theory, the changes in the energy levels of a Hydrogenlike atom produced by the increase of a unit in the charge of the nucleus, resulting from, for example, β decay.

By comparing the result with the exact one, discuss the validity of the approximation used.

Solution

Remember that the energy spectrum of a Hydrogenlike atom having atomic number Z is given (see problem 3.18) by

$$E_n^0 = -\frac{1}{2} mc^2 Z^2 \alpha^2 \frac{1}{n^2},$$

where $n = 1, 2, \ldots$ and $\alpha = e^2/\hbar c$ is the fine-structure constant. Due to the increase in charge, the potential energy is modified:

$$\mathcal{V} = -\frac{(Z+1)e^2}{r} = -\frac{Ze^2}{r} - \frac{e^2}{r} = \mathcal{V}_0 + \mathcal{V}_1.$$

Therefore, the first-order change in the energy levels is given by

$$E_n^1 = \langle n^0|\mathcal{V}_1|n^0\rangle = -e^2\left\langle \frac{1}{r} \right\rangle_{n^0}.$$

The $\frac{1}{r}$ expectation value can be calculated in many ways (see, e.g., problem 3.19). Here, we use the Virial Theorem for Coulomb potential, relating the expectation values of kinetic and potential energies $\langle \mathcal{T} \rangle = -\langle \mathcal{V} \rangle / 2$ (see problem 1.10). We get

$$E_n^0 = \langle \mathcal{T} + \mathcal{V}_0 \rangle_{n^0} = \frac{1}{2} \langle \mathcal{V}_0 \rangle_{n^0} = -\frac{1}{2} Z e^2 \left\langle \frac{1}{r} \right\rangle_{n^0}$$

and, therefore,

$$E_n^1 = \frac{2 E_n^0}{Z}.$$

This result is to be compared with the exact one,

$$E_n = -\frac{1}{2} m c^2 (Z + 1)^2 \alpha^2 \frac{1}{n^2} = -\frac{1}{2} m c^2 \alpha^2 \frac{1}{n^2} (Z^2 + 2Z + 1) = E_n^0 + E_n^1 - \frac{1}{2} m c^2 \alpha^2 \frac{1}{n^2},$$

which can be written in the form

$$E_n = E_n^0 \left(1 + \frac{2}{Z} - \frac{1}{Z^2} \right).$$

This expression shows that Perturbation Theory provides a good approximation if $Z \gg 1$.

6.24 Stark Effect

When a Hydrogen atom is placed in an electric field, its emission or absorption electromagnetic spectrum lines split into neighboring components. This phenomenon is called the Stark Effect and is attributed to the degeneracy breakdown due to the interaction between the electric field and the electric dipole moment of the atom.

Consider a Hydrogen atom in the presence of a constant electric field $\mathcal{E}$ directed along the z axis, giving rise to the interaction energy

$$\mathcal{H}_1 = e \mathcal{E} \cdot \mathbf{r} = e \mathcal{E} r \cos \theta.$$

Determine the first-order perturbative corrections in the two lower energy levels.

Solution

We recall that the eigenvalues and eigenfunctions of the Hydrogen atom Hamiltonian in the absence of external fields (see problem 3.18) are given by

$$E_n = -\frac{1}{2} m c^2 \frac{\alpha^2}{n^2}, \qquad \psi_{n,l,m}(\mathbf{r}) = R_{n,\ell}(r) Y_{\ell,m}(\theta, \phi), \tag{6.5}$$

where

$$\alpha = \frac{e^2}{\hbar c} \quad \text{and} \quad a_0 = \frac{\hbar^2}{m e^2}$$

are the fine-structure constant and the Bohr radius. In terms of the Laguerre polynomials and Legendre functions, the eigenfunctions are given by

$$\psi_{n,\ell,m}(\mathbf{r}) = N_{n,\ell}\, r^\ell\, e^{-\frac{r}{n a_0}} L_{n-\ell-1}^{2\ell+1}\left(\frac{2r}{n a_0}\right) P_\ell^m(\cos\theta)\, e^{\imath m\phi}, \tag{6.6}$$

where $N_{n,\ell}$ is a normalization constant.

Ground State

The first-order change in the ground state energy is given by

$$E_1^1 = e\,\langle 1, 0, 0|\, \mathbf{r}\, |1, 0, 0\rangle \cdot \mathcal{E},$$

which expresses the correction as the scalar product of the electric field times the average of the coordinate in the unperturbed state. In terms of eigenfunctions, we have

$$E_1^1 = e\,\mathcal{E} \cdot \int d\mathbf{r}\, |\psi_{1,0,0}(\mathbf{r})|^2\, \mathbf{r}.$$

This quantity is zero; more precisely, this is true for every diagonal matrix element:

$$\langle n, \ell, m|\, \mathbf{r}\, |n, \ell, m\rangle = \int d\mathbf{r}\, |\psi_{n,\ell,m}(\mathbf{r})|^2\, \mathbf{r} = 0.$$

In fact, $\psi_{n,\ell,m}$, which is the product of a function of r times a Spherical Harmonic, has the same parity as the latter, i.e., $(-1)^\ell$. Its square modulus has even parity and the integral is zero, because you have to integrate an odd function over the whole space. So, there is no first-order correction in the ground state energy.

First Excited State

The Stark Effect the result of the fact that the degeneracy of $n > 1$ levels does not survive in the presence of an external electric field.

Consider states with $n = 2$ in the absence of an electric field. There are four eigenfunctions (A.64) relative to the same eigenvalue $E_2 = \frac{E_1}{4}$, to which we associate four kets as follows:

$$|1\rangle \;\rightarrow\; \psi_{2,0,0}(r, \theta, \phi) = \frac{1}{4\sqrt{2\pi}}\, a_0^{-\frac{3}{2}}\left(2 - \frac{r}{a_0}\right) e^{-\frac{r}{2a_0}}, \tag{6.7}$$

$$|2\rangle \;\rightarrow\; \psi_{2,1,0}(r, \theta, \phi) = \frac{1}{4\sqrt{2\pi}}\, a_0^{-\frac{3}{2}}\, \frac{r}{a_0}\, e^{-\frac{r}{2a_0}} \cos\theta, \tag{6.8}$$

$$|3\rangle \; \rightarrow \; \psi_{2,1,1}(r, \theta, \phi) = \frac{1}{8\sqrt{2\pi}} \, a_0^{-\frac{3}{2}} \, \frac{r}{a_0} \, e^{-\frac{r}{2a_0}} \sin\theta \, e^{i\phi}, \tag{6.9}$$

$$|4\rangle \; \rightarrow \; \psi_{2,1,-1}(r, \theta, \phi) = \frac{1}{8\sqrt{2\pi}} \, a_0^{-\frac{3}{2}} \, \frac{r}{a_0} \, e^{-\frac{r}{2a_0}} \sin\theta \, e^{-i\phi}. \tag{6.10}$$

The first-order corrections are given by the eigenvalues of the matrix representative of $\mathcal{H}_1$ in the eigenspace of E_1. We can use the numbering used for the states to identify the elements of this matrix:

$$(\mathcal{H}_1)_{k,j} = \langle k|\mathcal{H}_1|j\rangle \quad \text{with } k, j = 1, 2, 3, 4.$$

As we have seen, the matrix diagonal elements are zero for parity reasons. Also, since it is worth (3.15), as found in problem 3.10,

$$\Delta m = m' - m = 0 \quad \text{and} \quad \Delta\ell = \ell' - \ell = \pm 1, \tag{6.11}$$

all of the matrix elements relative to couples of eigenkets having different L_z eigenvalues are null. Therefore, the matrix becomes

$$(\mathcal{H}_1)_{m,j} = \begin{pmatrix} 0 & \langle 1|\mathcal{H}_1|2\rangle & 0 & 0 \\ \langle 2|\mathcal{H}_1|1\rangle & 0 & 0 & 0 \\ 0 & 0 & 0 & 0 \\ 0 & 0 & 0 & 0 \end{pmatrix}.$$

Two eigenvalues are zero and the other two (the correction in the energy level) are obtained as eigenvalues of the 2×2 block in the $m = 0$ sector

$$\begin{pmatrix} 0 & \langle 1|\mathcal{H}_1|2\rangle \\ \langle 2|\mathcal{H}_1|1\rangle & 0 \end{pmatrix}. \tag{6.12}$$

As the matrix is hermitian, the non-zero matrix elements are complex conjugates, indeed, equal, being real. Let us calculate them:

$$\langle 1|\mathcal{H}_1|2\rangle = e\mathcal{E} \int r^2 dr \, d\cos\theta \, d\phi \, \psi_{2,0,0}(r, \theta, \phi) \, r\cos\theta \, \psi_{2,1,0}(r, \theta, \phi) =$$

$$= \frac{e\mathcal{E}}{16 \cdot 2\pi} \, a_0^{-3} \int_0^{2\pi} d\phi \int_{-1}^{1} d\cos\theta \, \cos^2\theta \int_0^{\infty} dr \, r^3 \frac{r}{a_0} \left(2 - \frac{r}{a_0}\right) e^{-\frac{r}{a_0}} =$$

$$= \frac{e\mathcal{E}}{24} \, a_0 \int_0^{\infty} dx \, x^4(2 - x) \, e^{-x} =$$

$$= -3 \, e \, \mathcal{E} a_0.$$

In the last step, integral (A.6) was used.

So, matrix (6.12), is given by

$$\begin{pmatrix} 0 & -3\,e\,\mathcal{E}\,a_0 \\ -3\,e\,\mathcal{E}\,a_0 & 0 \end{pmatrix}.$$

Its eigenvalues are easily calculated:

$$3\,e\,\mathcal{E}\,a_0 \quad \text{relative to the eigenvector} \quad \frac{1}{\sqrt{2}}\begin{pmatrix} 1 \\ -1 \end{pmatrix},$$

$$-3\,e\,\mathcal{E}\,a_0 \quad \text{relative to the eigenvector} \quad \frac{1}{\sqrt{2}}\begin{pmatrix} 1 \\ 1 \end{pmatrix}.$$

Summing up our results, level E_2, 4-times degenerate with eigenkets $|2, 0, 0\rangle$, $|2, 1, 0\rangle$, $|2, 1, 1\rangle$ e $|2, 1, -1\rangle$, separates into 3 levels at the first perturbative order:

$$E_2 + 3\,e\,\mathcal{E}\,a_0 \quad \text{not degenerate relative to the eigenket} \quad \frac{1}{\sqrt{2}}\,(|2, 0, 0\rangle - |2, 1, 0\rangle),$$
$$E_2 \quad \text{2 times degenerate relative to the eigenkets} \quad |2, 1, 1\rangle \text{ and } |2, 1, -1\rangle,$$
$$E_2 - 3\,e\,\mathcal{E}\,a_0 \quad \text{not degenerate relative to the eigenket} \quad \frac{1}{\sqrt{2}}\,(|2, 0, 0\rangle + |2, 1, 0\rangle).$$

6.25 Hydrogen: Relativistic Corrections

The expression currently used for kinetic energy $T = \frac{p^2}{2m}$ can be considered as an approximation for velocities that are small with respect to the speed of light c of the relativistic expression:

$$T = \sqrt{p^2 c^2 + m^2 c^4}.$$

Expanding this expression in the power series of $\frac{p}{c}$, we get

$$T = mc^2\sqrt{1 + \frac{p^2}{m^2 c^2}} \simeq mc^2 + \frac{p^2}{2m} - \frac{p^4}{8m^3 c^2},$$

where the terms of order above $(\frac{p}{c})^4$ were ignored. The first term is the rest energy and only contributes to redefining the arbitrary constant relative to energy.
Using results obtained in problem 3.19,

$$\left\langle \frac{e^2}{r} \right\rangle_{n,\ell,m} = -2E_n^0,$$

$$\left\langle \frac{e^4}{r^2} \right\rangle_{n,\ell,m} = 4(E_n^0)^2\,\frac{n}{\ell + \frac{1}{2}},$$

calculate the first-order Perturbation Theory effects on the energy spectrum derived from the term

$$\mathcal{H}_1 = -\frac{p^4}{8m^3c^2}. \tag{6.13}$$

Solution

We note that, since

$$p^4 = 4m^2 \left(\mathcal{H}_0 + \frac{e^2}{r}\right)^2$$

commutates with L^2 and L_z, the matrices of $\mathcal{H}_1$ in each eigenvalue eigenspace are diagonal, and we can use non-degenerate Perturbation Theory. Therefore, the first-order shifts in the energy levels are given by

$$E_n^1 = -\frac{1}{8m^3c^2} \langle n, \ell, m | p^4 | n, \ell, m \rangle =$$

$$= \frac{1}{2mc^2} \langle n, \ell, m | \left(\mathcal{H}_0 + \frac{e^2}{r}\right)^2 | n, \ell, m \rangle =$$

$$= -\frac{1}{2mc^2} \left((E_n^0)^2 + 2E_n^0 \langle n, \ell, m | \frac{e^2}{r} | n, \ell, m \rangle + \langle n, \ell, m | \frac{e^4}{r^2} | n, \ell, m \rangle \right).$$

By using the above expressions for the expectation values, the final result is obtained:

$$E_n^1 = -\frac{1}{2} m c^2 \alpha^4 \left[-\frac{3}{4n^4} + \frac{1}{n^3(\ell + \frac{1}{2})} \right], \tag{6.14}$$

showing that there is no more degeneracy in the quantum number ℓ.

6.26 Spin-Orbit Interaction

In the Hydrogen atom, the interaction of the magnetic moment of the electron and the magnetic field generated from its orbital motion around the nucleus leads to an extra term in the Hamiltonian:

$$\mathcal{H}_{SO} = -\mu_e \cdot \mathbf{B} = \frac{e^2}{2m^2c^2r^3} \mathbf{S} \cdot \mathbf{L},$$

known as the spin-orbit interaction.
Using the result (obtained in problem 3.19)

$$\left\langle \frac{1}{r^3} \right\rangle_{n,\ell} = \frac{1}{a_0^3} \frac{1}{n^3\ell(\ell + \frac{1}{2})(\ell + 1)} = \left(\frac{mc\alpha}{\hbar}\right)^3 \frac{1}{n^3\ell(\ell + \frac{1}{2})(\ell + 1)},$$

calculate the first-order perturbative corrections in the energy spectrum due to this interaction.

Solution

Notice that, having denoted the total angular momentum by **J**, it results that

$$\mathbf{S} \cdot \mathbf{L} = \frac{1}{2} \left[(\mathbf{S} + \mathbf{L})^2 - L^2 - S^2 \right] = \frac{1}{2} [J^2 - L^2 - S^2].$$

$\mathcal{H}_{SO}$ commutates, therefore, with the set of compatible variables J^2, L^2, S^2, and there is a basis common to these operators and the Hamiltonian, $|n, j, m_j; \ell; s\rangle$. In this basis, $\mathcal{H}_{SO}$ is represented by a diagonal matrix giving the required corrections:

$$\langle n', j', m'_j; \ell'; s' | \mathcal{H}_{SO} | n, j, m_j; \ell; s \rangle =$$

$$= \delta_{n,n'} \, \delta_{j,j'} \, \delta_{m,m'} \, \delta_{\ell,\ell'} \, \frac{e^2}{4m^2c^2} \left\langle \frac{1}{r^3} \right\rangle_{n,\ell} \hbar^2 \left[j(j+1) - \ell(\ell+1) - \frac{3}{4} \right].$$

The total angular momentum, the sum of the orbital angular momentum and the spin, has a quantum number j ranging from $|j - s|$ to $j + s$, $j = \ell \pm \frac{1}{2}$. Thus, we have

$$E_n^1 = \frac{\hbar^2 e^2}{4m^2c^2} \left\langle \frac{1}{r^3} \right\rangle_{n,\ell} \times \begin{cases} \ell & \text{if } j = \ell + \frac{1}{2} \\ -(\ell + 1) & \text{if } j = \ell - \frac{1}{2} \end{cases}.$$

Using the suggested expression for the expectation value of r^{-3}, we get the final result

$$E_n^1 = \frac{1}{4} mc^2 \alpha^4 \frac{1}{n^3 \ell(\ell + \frac{1}{2})(\ell + 1)} \times \begin{cases} \ell & \text{if } j = \ell + \frac{1}{2} \\ -(\ell + 1) & \text{if } j = \ell - \frac{1}{2} \end{cases} \tag{6.15}$$

6.27   Ground State of Helium

The Helium atom has atomic number $Z = 2$ and mass number $A = 4$; thus, there are two electrons moving around a nucleus of mass equal to 4 times that of Hydrogen and about 8×10^3 that of an electron. Assuming this mass to be infinite, we refer to the coordinates of the two electrons with respect to the nucleus with $\mathbf{r}_1$ and $\mathbf{r}_2$ and to their relative positions with with $\mathbf{r}_{12} = \mathbf{r}_1 - \mathbf{r}_2$. The Hamiltonian of the Helium atom is, therefore,

$$\mathcal{H} = -\frac{\hbar^2}{2m} (\nabla_1^2 + \nabla_2^2) - \frac{Ze^2}{r_1} - \frac{Ze^2}{r_2} + \frac{e^2}{r_{12}}.$$

If there were no repulsive interaction between the two electrons, in the Schrödinger equation, there would be the separation between the variables of the two electrons and, E being the energy of the ground state, we would have

$$\psi_E^0(\mathbf{r}_1, \mathbf{r}_2) = \psi_{E_1}(\mathbf{r}_1)\,\psi_{E_1}(\mathbf{r}_2) = \frac{Z^3}{\pi a_0^3}\, e^{-\frac{Z(r_1+r_2)}{a_0}}. \tag{6.16}$$

Here, we denoted by E_1 and $\psi_{E_1}(r) = (2/\sqrt{4\pi})(Z/a_0)^{\frac{3}{2}} \cdot e^{-\frac{Zr}{a_0}}$, respectively, the ground state energy of a Hydrogenlike atom with $Z = 2$ and the corresponding eigenfunction. In this very rough approximation, neglecting the positive term of attraction between the electrons, the ground state energy would be

$$E^0 = 2\,E_1 = 2\left[-\frac{m(Ze^2)^2}{2\hbar^2}\right] = -8 \cdot 13.6\,\mathrm{eV} = -108.8\,\mathrm{eV}.$$

Experimentally, it is found that the ground state of the Helium has energy $E = -78.98\,\mathrm{eV}$, a much higher value.

Consider the interaction between the two electrons as a perturbation and calculate the first-order change to E^0.

Solution

The Helium atom Hamiltonian can be written as

$$\mathcal{H} = \mathcal{H}_0 + \mathcal{H}_1,$$

where

$$\mathcal{H}_0 = -\frac{\hbar^2}{2m}(\nabla_1^2 + \nabla_2^2) - \frac{Ze^2}{r_1} - \frac{Ze^2}{r_2},$$

and the perturbation is given by

$$\mathcal{H}_1 = \frac{e^2}{r_{12}}.$$

The first-order correction due to the interaction between the two electrons is given by

$$E^1 = \langle \psi_E^0 | \mathcal{H}_1 | \psi_E^0 \rangle =$$

$$= \left(\frac{Z^3}{\pi a_0^3}\right)^2 \int d\mathbf{r}_1 d\mathbf{r}_2 e^{-\frac{2Z(r_1+r_2)}{a_0}} \frac{e^2}{r_{12}} =$$

$$= \left(\frac{Z^3 e}{\pi a_0^3}\right)^2 K.$$

Having introduced $\beta = \frac{2Z}{a_0}$, K is the following integral:

$$K = \int d\mathbf{r_1} d\mathbf{r_2} \, e^{-\beta(r_1+r_2)} \frac{1}{|\mathbf{r_1} - \mathbf{r_2}|} = \int d\mathbf{r_1} e^{-\beta r_1} f(r_1),$$

where

$$f(r_1) = \int d\mathbf{r_2} \, e^{-\beta r_2} \frac{1}{|\mathbf{r_1} - \mathbf{r_2}|}$$

is a function of $\mathbf{r_1}$ only, because, by integrating over the whole $\mathbf{r_2}$ solid angle, the dependence on the angles disappears. To calculate $f(r_1)$, we are free, therefore, to orient $\mathbf{r_1}$ in the direction of the z axis.

We get

$$f(r_1) = 2\pi \int dr_2 \, r_2^2 \, e^{-\beta r_2} \int_{-1}^{+1} d\cos\theta \, \frac{1}{\sqrt{r_1^2 + r_2^2 - 2r_1 r_2 \cos\theta}} =$$

$$= -2\pi \int dr_2 \, r_2^2 \, e^{-\beta r_2} \frac{1}{r_1 r_2} \left[\sqrt{r_1^2 + r_2^2 - 2r_1 r_2 \cos\theta} \right]_{-1}^{+1} =$$

$$= \frac{2\pi}{r_1} \int dr_2 \, r_2 \, e^{-\beta r_2} (r_1 + r_2 - |r_1 - r_2|) =$$

$$= \frac{2\pi}{r_1} \left[\int_0^{r_1} dr_2 \, r_2 \, e^{-\beta r_2} (2r_2) + \int_{r_1}^{\infty} dr_2 \, r_2 \, e^{-\beta r_2} (2r_1) \right] =$$

$$= \frac{4\pi}{r_1 \beta^3} \left[2 - e^{-\beta r_1} (2 + \beta r_1) \right],$$

where we used the expressions, derived from (A.7),

$$I_2(0, r_1) = \frac{1}{\beta^3} [2 - (2 + 2r_1\beta + r_1^2\beta^2)e^{-\beta r_1}]$$

and

$$I_1(r_1, \infty) = \frac{1}{\beta^2} (1 + \beta r_1)e^{-\beta r_1}.$$

Returning to the calculation of K, we obtain, using the usual integrals (A.8),

$$K = 4\pi \int_0^{\infty} dr_1 \, r_1^2 e^{-\beta r_1} \frac{4\pi}{r_1 \beta^3} \left[2 - e^{-\beta r_1} (2 + \beta r_1) \right] = \frac{(4\pi)^2}{\beta^5} \frac{5}{4}.$$

By inserting these results into the expression for E^1, we obtain

$$E^1 = \left(\frac{Z^3 e}{\pi a_0^3} \right)^2 \frac{5}{4} \frac{(4\pi)^2}{\beta^5} = \frac{5}{8} \frac{Ze^2}{a_0} = \frac{5}{4} Z \left(\frac{1}{2}mc^2\alpha^2 \right) = 34 \, \text{eV},$$

where it has been taken into account that $a_0 = \frac{\hbar}{mc\alpha}$, $\alpha = \frac{e^2}{\hbar c}$ and that $\frac{1}{2}mc^2\alpha^2 = 13, 6\,\text{eV}$ is the energy, changed in sign, of the ground state of the Hydrogen atom. In this approximation, the energy of the ground state of the Helium atom is

$$E = E^0 + E^1 = -108.8\,\text{eV} + 34\,\text{eV} = -74, 8\,\text{eV}.$$

This value is very close to the previously mentioned experimental value $-78.98\,\text{eV}$. An even better approximation is obtained using the variational method (see problem 11.4).

Chapter 7
Time-Dependent Perturbation Theory

7.1 Harmonic Oscillator: Instantaneous Perturbation

Two particles of mass m move along the x axis, interacting by means of a force having characteristic elastic constant k.

Assuming that, while they are in the ground state of energy E_0, the constant k is suddenly halved, what is the probability that a new energy measure will result in the energy of the ground state?

Solution

Having denoted the two particles' coordinates by x_1 and x_2, we introduce the center of mass and relative coordinates

$$X = \frac{x_1 + x_2}{2}, \quad x = x_1 - x_2,$$

the total mass $M = 2m$ and the reduced mass $\mu = \frac{m}{2}$. Placing

$$\Psi(X, x) = \Phi(X)\psi(x),$$

the Schrödinger equation

$$\left[-\frac{\hbar^2}{2m} \frac{\partial^2}{\partial x_1^2} - \frac{\hbar^2}{2m} \frac{\partial^2}{\partial x_2^2} + V(x_1 - x_2) - W \right] \Psi(x_1, x_2) = 0$$

separates into two equations

$$\left[-\frac{\hbar^2}{2M} \frac{d^2}{dX^2} - E_{CM} \right] \Phi(X) = 0,$$

$$\left[-\frac{\hbar^2}{2\mu} \frac{d^2}{dx^2} + V(x) - E \right] \psi(x) = 0, \tag{7.1}$$

© Springer Nature Switzerland AG 2019
L. Angelini, *Solved Problems in Quantum Mechanics*, UNITEXT for Physics,
https://doi.org/10.1007/978-3-030-18404-9_7

satisfying the condition

$$W = E_{CM} + E .$$

The center of mass moves freely, while the reduced mass is subject to a harmonic potential. For the purpose of the problem, we are only interested in the relative coordinate motion.

Initially, the system is described by the wave function (A.16)

$$\psi_0(x) = \left(\frac{\alpha}{\sqrt{\pi}}\right)^{\frac{1}{2}} e^{-\frac{1}{2}\alpha^2 x^2} . \quad \text{where} \quad \alpha = \sqrt{\frac{m\omega}{\hbar}} = \sqrt[4]{\frac{mk}{\hbar^2}} .$$

We want to know the probability that, after the halving of the elastic constant, the oscillator will be in the ground state of the new system, that is, in the state described by

$$\psi_0'(x) = \left(\frac{\alpha'}{\sqrt{\pi}}\right)^{\frac{1}{2}} e^{-\frac{1}{2}\alpha'^2 x^2} , \quad \text{where} \quad \alpha' = \sqrt{\frac{m\omega'}{\hbar}} = \sqrt[4]{\frac{mk'}{\hbar^2}} = \frac{\alpha}{\sqrt[4]{2}} .$$

Since it is an instantaneous perturbation, the state of the particle does not change, but its Hamiltonian does change. Therefore, the required probability is given by the square modulus of

$$\langle \psi_0 | \psi_0' \rangle = \int_{-\infty}^{+\infty} dx \sqrt{\frac{\alpha\alpha'}{\pi}} e^{-\frac{1}{2}(\alpha^2 + \alpha'^2)x^2} =$$

$$= \sqrt{\frac{\alpha\alpha'}{\pi}} \sqrt{\frac{2\pi}{\alpha^2 + \alpha'^2}} = \sqrt{\frac{2\alpha\alpha'}{\alpha^2 + \alpha'^2}} ,$$

where we used (A.1). So, the required probability is given by

$$P_0 = \frac{2\alpha\alpha'}{\alpha^2 + \alpha'^2} = \frac{2\sqrt{2}}{\sqrt[4]{2}(1 + \sqrt{2})} = 2^{\frac{5}{4}}(\sqrt{2} - 1) = 0.9852 .$$

7.2 Harmonic Oscillator in an Electric Field: Instantaneous Perturbation

A particle of mass m and charge q is subject to harmonic oscillations of frequency ω along the x axis. It is suddenly placed in a uniform electric field that generates a potential

$$\mathcal{H}_1 = -q\mathcal{E}X .$$

Determine the transition probabilities in the case in which the system is initially in the ground state. Use first-order perturbation theory to approximate these probabilities in the case of a weak electric field.

Solution

After the electric field is switched on, the Hamiltonian becomes

$$\mathcal{H} = \mathcal{H}_0 + \mathcal{H}_1 = \frac{p^2}{2m} + \frac{1}{2}m\omega^2 x^2 - q\mathcal{E}X = \frac{p^2}{2m} + \frac{1}{2}m\omega^2 (x - x_0)^2 - \frac{1}{2}\frac{q^2\mathcal{E}^2}{m\omega^2},$$

which still represents the Hamiltonian of a harmonic oscillator with the same frequency, but with energy levels translated by the quantity

$$-\frac{1}{2}\frac{q^2\mathcal{E}^2}{m\omega^2}$$

and center of oscillation placed in

$$x_0 = \frac{q\mathcal{E}}{m\omega^2}.$$

$\mathcal{H}_0$ eigenfunctions are given by (A.16)

$$\phi_n^0(x) = \left(\frac{m\omega}{\pi\hbar}\right)^{\frac{1}{4}} \frac{1}{\sqrt{2^n n!}} e^{-\frac{\xi^2}{2}} H_n(\xi), \qquad \xi = \sqrt{\frac{m\omega}{\hbar}}x$$

whereas, having denoted the eigenkets of the new Hamiltonian $\mathcal{H}$ by $|k\rangle$, its eigenfunctions are

$$\langle x|k\rangle = \phi_k(x) = \phi_k^0(x - x_0) = \left(\frac{m\omega}{\pi\hbar}\right)^{\frac{1}{4}} \frac{1}{\sqrt{2^k k!}} e^{-\frac{(\xi-\xi_0)^2}{2}} H_k(\xi - \xi_0)$$

with

$$\xi_0 = \sqrt{\frac{m\omega}{\hbar}}x_0.$$

Remembering that (A.17)

$$H_n(\xi) = (-1)^n e^{\xi^2} \frac{d^n e^{-\xi^2}}{d\xi^n},$$

the transition amplitudes are given by

$$\langle k|0^0\rangle = \int dx\, \phi_k^*(x)\, \phi_0^0(x) = \int dx\, \phi_k^0(x - x^0)\, \phi_0^0(x) =$$

$$= \left(\frac{m\omega}{\pi\hbar}\right)^{\frac{1}{2}} \frac{1}{\sqrt{2^k k!}} (-1)^k \int_{-\infty}^{\infty} dx\, e^{-\frac{(\xi-\xi_0)^2}{2}} e^{(\xi-\xi_0)^2} \frac{d^k\, e^{-(\xi-\xi_0)}}{d\xi^k} e^{-\frac{\xi^2}{2}} =$$

$$= \frac{(-1)^k}{(\pi 2^k k!)^{\frac{1}{2}}} e^{-\frac{\xi_0^2}{2}} \int_{-\infty}^{\infty} d\xi\, e^{-\xi\xi_0} \frac{d^k\, e^{-\xi^2+2\xi\xi_0}}{d\xi^k}. \tag{7.2}$$

Using integration by parts, we get

$$
\int_{-\infty}^{\infty} d\xi\, e^{-\xi\xi_0}\, \frac{d^k\, e^{-\xi^2+2\xi\xi_0}}{d\xi^k} =
$$

$$
= e^{-\xi\xi_0}\, \frac{d^{k-1} e^{-\xi^2+2\xi\xi_0}}{d\xi^{k-1}}\Bigg|_{-\infty}^{\infty} - \int_{-\infty}^{\infty} d\xi\,(-\xi_0)\, e^{-\xi\xi_0}\, \frac{d^{k-1} e^{-\xi^2+2\xi\xi_0}}{d\xi^{k-1}}. \quad (7.3)
$$

The first term is null, and, after iterating k times the integration by parts, (7.2) becomes

$$
\langle k|0^0\rangle = \frac{(-1)^k}{(\pi 2^k k!)^{\frac{1}{2}}}\, e^{-\frac{\xi_0^2}{2}}\, \xi_0^k \int_{-\infty}^{\infty} d\xi\, e^{-\xi^2+\xi\xi_0} =
$$

$$
= \frac{(-1)^k}{(\pi 2^k k!)^{\frac{1}{2}}}\, e^{-\frac{\xi_0^2}{2}}\, \xi_0^k\, e^{\frac{\xi_0^2}{4}}\, \sqrt{\pi} = \frac{(-1)^k}{(2^k k!)^{\frac{1}{2}}}\, e^{-\frac{\xi_0^2}{4}}\, \xi_0^k. \quad (7.4)
$$

The transition probabilities are obtained by squaring the amplitudes

$$
P_{0,k} = \left(\frac{\xi_0^2}{2}\right)^k \frac{e^{-\frac{\xi_0^2}{2}}}{k!}.
$$

As a function of k, this is a Poisson distribution with expectation value $\frac{\xi_0^2}{2}$. Since

$$
\xi_0 = \sqrt{\frac{m\omega}{\hbar}}\, x_0 = \sqrt{\frac{m\omega}{\hbar}}\, \frac{q\mathcal{E}}{m\omega^2} = \frac{q\mathcal{E}}{\sqrt{m\hbar\omega^3}},
$$

if the perturbation is small, ξ_0 is also small, and the only major probability of transition to a state with $k \neq 0$ is for $k = 1$:

$$
P_{0,1} = \frac{\xi_0^2}{2}\, \frac{e^{-\frac{\xi_0^2}{2}}}{1!} \simeq \frac{\xi_0^2}{2} = \frac{q^2\mathcal{E}^2}{2m\hbar\omega^3}.
$$

This result is confirmed by the application of Perturbation Theory (A.80)

$$
P_{0,k} = \left|\frac{\langle k^0|\mathcal{H}_1|n^0\rangle}{E_k^0 - E_n^0}\right|^2 = \frac{q^2\mathcal{E}^2}{(k\hbar\omega)^2}\, |\langle k^0|\mathcal{H}_1|n^0\rangle|^2 =
$$

$$
= \frac{q^2\mathcal{E}^2}{(k\hbar\omega)^2}\, \left|\sqrt{\frac{\hbar}{2m\omega}}\langle k^0|a + a^\dagger|n^0\rangle\right|^2 = \frac{q^2\mathcal{E}^2}{(k\hbar\omega)^2}\, \frac{\hbar}{2m\omega}\delta_{k,1} = \frac{q^2\mathcal{E}^2}{2m\hbar\omega^3}\delta_{k,1}.
$$

7.3 Particle Confined on a Segment: Square Perturbation

Consider a particle of mass m bound to move along a segment of length a.

(a) Write the first 4 eigenfunctions and the corresponding eigenvalues.
(b) Suppose that the particle is in the ground state ($n = 1$). At time $t = 0$, it is instantaneously switched on a potential square of depth $-V_0$ ($V_0 > 0$) and width $b \ll a$ centered around $x = \frac{a}{2}$. If this potential is removed after a time Δt, what will be the probability of finding the system in each of the states with $n = 2$, $n = 3$ and $n = 4$?

Solution

(a) The required eigenvalues and eigenfunctions are given by

$$
E_1 = \frac{\hbar^2 \pi^2}{2ma^2}, \quad \psi_1(x) = \sqrt{\frac{2}{a}} \sin \frac{\pi x}{a},
$$

$$
E_2 = \frac{4\hbar^2 \pi^2}{2ma^2}, \quad \psi_2(x) = \sqrt{\frac{2}{a}} \sin \frac{2\pi x}{a},
$$

$$
E_3 = \frac{9\hbar^2 \pi^2}{2ma^2}, \quad \psi_3(x) = \sqrt{\frac{2}{a}} \sin \frac{3\pi x}{a},
$$

$$
E_4 = \frac{16\hbar^2 \pi^2}{2ma^2}, \quad \psi_4(x) = \sqrt{\frac{2}{a}} \sin \frac{4\pi x}{a}.
$$

(b) The transition probabilities (A.82) are:

$$
P_{1\to n}(\Delta t) = \left| -\frac{i}{\hbar} \int_0^{\Delta t} dt \langle n|\mathcal{H}_1|1\rangle e^{i\omega_{n,1}t} \right|^2 = \frac{1}{\hbar^2} |\langle n|\mathcal{H}_1|1\rangle|^2 \left| \frac{e^{i\omega_{n,1}\Delta t} - 1}{i\omega_{n,1}} \right|^2,
$$

where the matrix elements are given by:

$$
\langle n|\mathcal{H}^1|1\rangle = -\frac{2V_0}{a} \int_{\frac{a-b}{2}}^{\frac{a+b}{2}} dx \sin \frac{n\pi x}{a} \sin \frac{\pi x}{a}
$$

and $\omega_{n,1} = \frac{E_n - E_1}{\hbar} = \frac{\hbar \pi^2}{2ma^2}(n^2 - 1)$ with $n \neq 1$.
This integral is zero for $n = 2$ and $n = 4$, because the integrand is the product of an odd function (the nth eigenfunction) times an even function (the ground state eigenfunction).
It remains to calculate the probability of a transition to the third level. Since

$$
\int_{\frac{a-b}{2}}^{\frac{a+b}{2}} dx \sin \frac{n\pi x}{a} \sin \frac{\pi x}{a} = -\frac{a\left(2\sin \frac{\pi b}{a} + \sin \frac{2\pi b}{a}\right)}{4\pi},
$$

the result is

$$P_{1\to3}(\Delta t) = \left(\frac{V_0}{\hbar\omega_{3,1}}\right)^2 \sin^2 \frac{\omega_{3,1}\,\Delta t}{2} \left(\frac{2\sin\frac{\pi b}{a} + \sin\frac{2\pi b}{a}}{\pi}\right)^2 .$$

Wanting to shorten the calculations, the integral can be approximated by the product of the integrand value at the center of the integration interval ($\frac{a}{2}$) times the width of the interval ($\frac{b}{2}$):

$$P_{1\to3}(\Delta t) \simeq \left(\frac{2V_0 b}{a}\right)^2 \frac{4}{\hbar^2\omega_{3,1}^2} \sin^2 \frac{\omega_{3,1}\,\Delta t}{2}.$$

7.4 Harmonic Oscillator: Gaussian Perturbation

Consider a one-dimensional harmonic oscillator in the ground state at time $t = -\infty$ in the presence of the perturbation

$$\mathcal{H}_1(t) = -qEX\,e^{-\frac{t^2}{\tau^2}}.$$

What is the probability of finding, at time $t = +\infty$, the oscillator in state $|n\rangle$ in first-order Perturbation Theory?

Solution

The probability amplitudes at time $t = +\infty$ are the coefficients of the expansion of the state ket in the energy basis (A.81, A.82):

$$d_n(\infty) = -\frac{\imath}{\hbar} \int_{-\infty}^{+\infty} dt\,(-qE)\langle n|X|0\rangle e^{-\frac{t^2}{\tau^2}} e^{\imath n\omega t}.$$

Since

$$X = \sqrt{\frac{\hbar}{2m\omega}}\,(a + a^\dagger),$$

and we know that $a^\dagger|0\rangle = |1\rangle$ and $a|0\rangle = 0$, the only non-zero coefficient is d_1:

$$d_1(\infty) = \frac{\imath qE}{\hbar}\sqrt{\frac{\hbar}{2m\omega}} \int_{-\infty}^{+\infty} dt\,e^{-\frac{t^2}{\tau^2}} e^{\imath\omega t} =$$

$$= \frac{\imath qE}{\hbar}\sqrt{\frac{\hbar}{2m\omega}}\sqrt{\pi\tau^2}\,e^{-\frac{\omega^2\tau^2}{4}}.$$

The desired probability of a transition is

$$P_{0\to1} = \frac{q^2 E^2 \pi \tau^2}{2m\omega\hbar} e^{-\frac{\omega^2 \tau^2}{2}}.$$

7.5 Harmonic Oscillator: Damped Perturbation

Starting from the instant $t = 0$, the perturbation

$$\mathcal{H}_1(x, t) = Ax^2 e^{-bt},$$

acts on a harmonic oscillator of mass m and frequency ω. Using first-order Perturbation Theory, determine the probabilities of transition from the ground state to the nth state after a long period of time.

Solution

The desired probability, for $n \neq 0$, is given by

$$P_{0\to n} = \frac{1}{\hbar^2} \left| \int_0^\infty dt \, \langle n|AX^2 e^{-bt}|0\rangle e^{\iota\omega_{n,0}t} \right|^2 =$$

$$= \frac{A^2}{\hbar^2} |\langle n|X^2|0\rangle|^2 \left| \int_0^\infty dt \, e^{(\iota\omega_{n,0}-b)t} \right|^2 =$$

$$= \frac{A^2}{\hbar^2} \frac{1}{\omega_{n,0}^2 + b^2} \left| \frac{\hbar}{2m\omega} (\sqrt{2}\,\delta_{n,2} + \delta_{n,0}) \right|^2,$$

where we used the result of problem (2.17). Therefore, the only possible transition is the one to state $n = 2$, having probability

$$P_{0\to2} = \frac{2A^2}{\hbar^2} \frac{1}{4\omega^2 + b^2} \left(\frac{\hbar}{2m\omega}\right)^2 = \frac{A^2}{2m^2\omega^2} \frac{1}{4\omega^2 + b^2}.$$

7.6 Hydrogen Atom in a Pulsed Electric Field

At time $t = -\infty$, a Hydrogen atom is in the ground state; an electric field is applied along the z axis:

$$\mathbf{E}(t) = E_0 \, e^{-\frac{t^2}{\tau^2}} \, \hat{\mathbf{k}}.$$

Determine, by first-order Perturbative Theory, the probability that, at time $t = +\infty$, the atom is in one of the $n = 2$ states.

Solution

Having denoted by $\mathbf{r}_1$ and $\mathbf{r}_2$, respectively, the position of the electron and that of the proton and by $V(\mathbf{r})$ the potential generated by the electric field on a charge placed in $\mathbf{r}$, we have

$$V(\mathbf{r}) = -\int \mathbf{E} \cdot d\mathbf{r} = -E_0\, z\, e^{-\frac{t^2}{\tau^2}}.$$

The potential energy resulting from the interaction of the external field with the atom dipole is therefore

$$\mathcal{H}_1(\mathbf{r}, t) = -eV(\mathbf{r}_1) + eV(\mathbf{r}_2) = ezE_0\, e^{-\frac{t^2}{\tau^2}},$$

where z is the component of $\mathbf{r} = \mathbf{r}_1 - \mathbf{r}_2$ in the field direction.

The states with $n = 2$ are 4: they are the state $\ell = 0$ and the 3 states $\ell = 1$, $m = 0, \pm 1$. The desired probability is given (A.82) by

$$P_{1\to 2} = \sum_{\ell=0,1} \sum_{m=-\ell,\ell} \left| d_{2,\ell,m}(+\infty) \right|^2,$$

where

$$d_{2,\ell,m}(+\infty) = -\frac{i}{\hbar} \int_{-\infty}^{+\infty} dt\, e^{i\omega_{2,1}t} \langle 2, \ell, m | \mathcal{H}_1 | 1, 0, 0 \rangle. \tag{7.5}$$

Having denoted by μ the Hydrogen atom reduced mass and by α the fine-structure constant, the transition frequency is the same for all $n = 2$ states:

$$\omega_{2,1} = -\frac{1}{2\hbar}\mu c^2 \alpha^2 \left(\frac{1}{2^2} - \frac{1}{1^2} \right) = \frac{3}{8\hbar}\mu c^2 \alpha^2.$$

To calculate the perturbation matrix elements, we note that $\mathcal{H}_1$ commutes with L_z, the operator whose quantum numbers label the unperturbed states. This means that only the matrix elements between states with the same value of m (selection rule $\Delta m = 0$ for the electric dipole transitions[1]) can be different from zero. Indeed,

$$[\mathcal{H}_1, L_z] = 0 \Rightarrow \langle n, \ell, m | [\mathcal{H}_1, L_z] | n', \ell', m' \rangle = \hbar(m' - m)\, \langle n, \ell, m | \mathcal{H}_1 | n', \ell', m' \rangle = 0.$$

We, therefore, are down to calculating only the two terms relative to $m = 0$. The other selection rule for dipole transition is $\Delta\ell = \pm 1$. Apart from the general rule, it is easy to convince oneself that the matrix element in (7.5) will be zero if ℓ does not change: we should integrate, over the whole space, an odd function, z, coming from $\mathcal{H}_1$, times the product of two eigenfunctions with the same parity. Only one non-zero matrix element remains to be calculated. Taking into account the expressions for the Hydrogen atom wave functions (A.64), we obtain:

---

[1] Electric dipole selection rules are calculated in Problem 3.10.

$$\langle 2,1,0|Z|1,0,0 \rangle =$$

$$= \frac{1}{4\sqrt{2\pi}} \frac{1}{\sqrt{\pi}} a_0^{-3} \int_0^\infty r^2 dr \int_{-1}^{+1} d\cos\theta \int_0^{2\pi} d\varphi \, \frac{r}{a_0} e^{-\frac{r}{2a_0}} \cos\theta \, r \cos\theta \, e^{-\frac{r}{a_0}} =$$

$$= \frac{1}{2\sqrt{2}} a_0^{-4} \int_0^\infty dr \, r^4 \, e^{-\frac{3r}{2a_0}} \int_{-1}^{+1} d\cos\theta \, \cos^2\theta =$$

$$= \frac{1}{3\sqrt{2}} a_0 \frac{2^5}{3^5} \int_0^\infty dx \, x^4 \, e^{-x} =$$

$$= \frac{2^7\sqrt{2}}{3^5} a_0,$$

where formula (A.6) has been used. By replacing this expression in the probability amplitude, we have:

$$d_{2,\ell,m}(+\infty) = -\frac{i}{\hbar} \int_{-\infty}^{+\infty} dt \, e^{i\omega_{2,1}t} \, e \, E_0 \frac{2^7\sqrt{2}}{3^5} a_0 \, e^{-\frac{t^2}{\tau^2}} =$$

$$= -\frac{i}{\hbar} e \, E_0 \, a_0 \frac{2^7\sqrt{2}}{3^5} e^{-\frac{(\omega_{2,1}\tau)^2}{4}} \int_{-\infty}^{+\infty} dt \, e^{-\frac{1}{\tau^2}(t-i\frac{\omega_{2,1}\tau^2}{4})^2} =$$

$$= -\frac{i}{\hbar} e \, E_0 \, a_0 \frac{2^7\sqrt{2}}{3^5} \tau \sqrt{\pi} \, e^{-\frac{(\omega_{2,1}\tau)^2}{4}}.$$

The requested probability is therefore given by

$$P_{1\to2} = \frac{\pi}{\hbar^2} (e \, E_0 \, a_0 \, \tau)^2 \frac{2^{15}}{3^{10}} e^{-\frac{(\omega_{2,1}\tau)^2}{2}}.$$

Chapter 8
Identical Particles

8.1 Two Fermions in a Potential Well

Two non-identical particles, both of spin $\frac{1}{2}$ and mass m, are forced to move along a segment of length L interacting through a potential

$$V = k\, \mathbf{S}_1 \cdot \mathbf{S}_2.$$

(a) Calculate the Hamiltonian's eigenvalues and eigenfunctions.
(b) What would change if the particles were identical?

Solution

(a) Inasmuch as the potential depends only on the spin states and the particles are not identical, we look for eigenfunctions factorized in the form

$$\Psi(1, 2) = \psi_{n_1,n_2}(x_1, x_2)\, \chi(\mathbf{S}_1, \mathbf{S}_2),$$

where

$$\psi_{n_1,n_2}(x_1, x_2) = \psi_{n_1}(x_1)\, \psi_{n_2}(x_2).$$

$\psi_n(x)$, con $n = 1, 2, \ldots$, are energy eigenfunctions for a particle in a potential well and $\chi(\mathbf{S}_1, \mathbf{S}_2)$ represents spin eigenstates. The potential depends only on the scalar product among the spins and can be rewritten in terms of the total spin $\mathbf{S} = \mathbf{S}_1 + \mathbf{S}_2$:

$$V = k\, \mathbf{S}_1 \cdot \mathbf{S}_2 = \frac{k}{2}\left[(\mathbf{S}_1 + \mathbf{S}_2)^2 - S_1^2 - S_2^2\right] = \frac{k}{2}\left[S^2 - \frac{3}{2}\hbar^2\right].$$

So, factoring occurs if we use the eigenstates of the total spin (S^2 and S_z), that is, the singlet states ($S = 0$) and triplet states ($S = 1$).
We conclude that the eigenfunctions common to the Hamiltonian, S^2 and S_z, are

© Springer Nature Switzerland AG 2019
L. Angelini, *Solved Problems in Quantum Mechanics*, UNITEXT for Physics,
https://doi.org/10.1007/978-3-030-18404-9_8

$$\Psi_{n_1,n_2;s,m_s}(x_1, x_2) = \psi_{n_1,n_2}(x_1, x_2)\, \chi_{s,m_s},$$

with

$$\chi_{0,0} = \frac{1}{\sqrt{2}}\left[\chi_+(1)\,\chi_-(2) - \chi_-(1)\,\chi_+(2)\right],$$

$$\chi_{1,-1} = \chi_-(1)\,\chi_-(2),$$

$$\chi_{1,0} = \frac{1}{\sqrt{2}}\left[\chi_+(1)\,\chi_-(2) + \chi_-(1)\,\chi_+(2)\right],$$

$$\chi_{1,+1} = \chi_+(1)\,\chi_+(2),$$

and they correspond to the energy eigenvalues

$$E_{n,s} = \frac{\pi^2\hbar^2}{2mL^2}\,n^2 + \frac{k}{2}\left[s(s+1) - \frac{3}{2}\right]\hbar^2, \quad \text{with } n^2 = n_1^2 + n_2^2 \text{ and } n_1, n_2 = 1, 2, \ldots$$

The degeneracy is the product of $(2s+1)$ for the degeneracy that occurs if there is more than one pair (n_1, n_2) leading to the same value of n.

(b) If the two particles are identical, the eigenvalues do not change, but their degeneracy is reduced. In fact, it is necessary to construct the symmetric and antisymmetric combinations of the eigenfunctions relative to the spatial coordinates (the spin eigenstates already have determinate symmetry properties) and to impose the antisymmetry of the overall self-functions.

In the case $S = 0$, the eigenvalues are $E_{n,0} = \frac{\pi^2\hbar^2}{2mL^2}\,n^2 - \frac{3}{4}\,k\hbar^2$ and the eigenfunction's spatial part must be symmetric. So, we have

$$\Psi_{n_1,n_2;0,0}(x_1, x_2) = \frac{1}{\sqrt{2}}\left[\psi_{n_1}(x_1)\,\psi_{n_2}(x_2) + \psi_{n_2}(x_1)\,\psi_{n_1}(x_2)\right]\chi_{0,0} \quad \text{if } n_1 \neq n_2$$

and

$$\Psi_{n_1,n_2;0,0}(x_1, x_2) = \psi_{n_1}(x_1)\,\psi_{n_2}(x_2)\,\chi_{0,0} \quad \text{if } n_1 = n_2.$$

States $S = 1$ correspond to eigenvalues $E_{n,1} = \frac{\pi^2\hbar^2}{2mL^2}\,n^2 + \frac{1}{4}\,k\hbar^2$, and their eigenfunction's spatial part must be antisymmetric: as a consequence, it must be $n_1 \neq n_2$. Their eigenfunctions are

$$\Psi_{n_1,n_2;1,m_s}(x_1, x_2) = \frac{1}{\sqrt{2}}\left[\psi_{n_1}(x_1)\,\psi_{n_2}(x_2) - \psi_{n_2}(x_1)\,\psi_{n_1}(x_2)\right]\chi_{1,m_s},$$

with $m_s = 0, \pm 1$.

8.2 Two Fermions in a Potential Well in the Presence of δ Potential

Two identical fermions of mass m, constrained to move along a segment of length L, have a spin component along the z axis equal to $+\frac{\hbar}{2}$.

(a) What are the minimum system energy and the corresponding eigenfunction?
(b) If there is an interaction potential $k\ \delta(x_1 - x_2)$, how does the value of the energy change to first-order Perturbation Theory?

Solution

(a) The system is composed of two identical non-interacting fermions. Its spin eigen-function is

$$\chi(s_{1,z}, s_{2,z}) = \chi\left(+\frac{\hbar}{2}, +\frac{\hbar}{2}\right).$$

Each of the two particles, if isolated, has energy eigenfunctions and eigenvalues given by

$$\psi_n(x) = \sqrt{\frac{2}{L}}\ \sin\frac{n\pi x}{L} \qquad E_n = \frac{n^2\pi^2\hbar^2}{2mL^2}, \qquad \text{con } n = 1, 2, \ldots$$

The two particles are non-interacting and, therefore, the spatial part of the generic eigenfunction can be written as a product of two eigenfunctions relative to the single-particle Hamiltonian and the eigenvalue is the sum of the two correspond-ing energy eigenvalues. However, since the overall wave function of the system must be antisymmetric in the exchange of the two particles and the spin function is symmetric, the spatial part must be antisymmetric. For this reason, the state of minimum energy cannot be that in which the two particles are in the ground state, but rather

$$\Psi_{s_{1,z},s_{2,z}}(x_1, x_2) = \frac{1}{\sqrt{2}}\left[\psi_1(x_1)\psi_2(x_2) - \psi_2(x_1)\psi_1(x_2)\right] \chi\left(+\frac{\hbar}{2}, +\frac{\hbar}{2}\right),$$

corresponding to the energy level

$$E_{1,2} = \frac{5\pi^2\hbar^2}{2mL^2}.$$

(b) Having introduced the δ interaction potential, the energy level is not modified by first-order Perturbation Theory. In fact, it results that

$$E'_{1,2} = k \int_0^L dx_1 \int_0^L dx_2 \, \Psi^*_{s_{1,z},s_{2,z}}(x_1, x_2) \, \delta(x_1 - x_2) \, \Psi_{s_{1,z},s_{2,z}}(x_1, x_2) =$$

$$= k \int_0^L dx \, \left| \Psi_{s_{1,z},s_{2,z}}(x, x) \right|^2 = 0,$$

due to the antisymmetry of the integrand function.

8.3   Two Interacting Fermions

Two particles of equal mass and spin $\frac{1}{2}$ are constrained to move along a line and interact via a potential

$$V = \frac{1}{2} kx^2 + a \, \mathbf{S}_1 \cdot \mathbf{S}_2,$$

where k and a are constants $(k > 0)$, x is the relative distance between the particles and $\mathbf{S}_1$ and $\mathbf{S}_2$ are their spin operators.

Determine the energy eigenvalues in the case in which the particles are different and that in which they are identical.

Solution

Once we have separated the motion of the center of mass from that of the relative coordinate x, we limit ourselves to considering the latter. We note that ($\mathbf{S}$ is the toral spin)

$$\mathbf{S}_1 \cdot \mathbf{S}_2 = \frac{1}{2} (S^2 - S_1^2 - S_2^2) = \frac{1}{2} (S^2 - \frac{3}{2} \hbar^2).$$

It follows that the Hamiltonian commutes with S^2 and S_z and, therefore, that we can consider the eigenstates common to $\mathcal{H}$, S^2 e S_z. They are given by the tensor product of the eigenkets of the harmonic oscillator Hamiltonian with the eigenkets common to S^2 e S_z.

The eigenvalues, if the particles are distinguishable, are therefore given by:

$$E_n^{s=1} = (n + \frac{1}{2}) \hbar\omega + \frac{1}{4} a\hbar^2 \qquad \text{for triplet states,}$$

$$E_n^{s=0} = (n + \frac{1}{2}) \hbar\omega - \frac{3}{4} a\hbar^2 \qquad \text{for singlet state,}$$

with $n = 0, 1, \ldots$.

Let us now consider the case of identical particles. Notice that changing the relative coordinate x in $-x$ is equivalent to the exchange of the two particles and that the nth eigenfunction of the harmonic oscillator Hamiltonian has parity of $(-1)^n$. As a consequence, the eigenvalues do not change, but the triplet states (symmetric) may exist only if n is odd and the singlet states (antisymmetric) only if n is even.

8.4 Two Identical Fermionic Oscillators

A system composed of two identical particles of spin $\frac{1}{2}$ constrained to move along a line is described by the Hamiltonian

$$\mathcal{H} = \frac{1}{2m}(p_1^2 + p_2^2) + \frac{1}{2}m\omega^2(x_1^2 + x_2^2).$$

Determine the complete wave function (spatial part and spin part) of the states corresponding to the ground level and the first excited level, as well as the related eigenvalues of $\mathcal{H}$, $\mathbf{S}^2 = (\mathbf{S}_1 + \mathbf{S}_2)^2$ and S_z.

Solution

For the ith particle, we denote denote the energy eigenfunctions for a one-dimensional harmonic oscillator by $\psi_n(x_i)$ and the spin eigenstates by $\chi_\pm(i)$. Being identical fermions, the eigenfunctions common to $\mathcal{H}$, S^2, S_z must be antisymmetric under the exchange of the two particles. As for the spatial coordinates, we can construct the following complete energy eigenfunctions having definite symmetry properties (indicated with $+$ and $-$):

$$\psi_0^+(x_1, x_2) = \psi_0(x_1)\,\psi_0(x_2) \qquad \text{with energy } E = \hbar\omega$$

and

$$\left.\begin{aligned}
\psi_1^+(x_1, x_2) &= \tfrac{1}{\sqrt{2}}\left[\psi_0(x1)\,\psi_1(x_2) + \psi_1(x1)\,\psi_0(x_2)\right] \\[4pt]
\psi_1^-(x_1, x_2) &= \tfrac{1}{\sqrt{2}}\left[\psi_0(x1)\,\psi_1(x_2) - \psi_1(x1)\,\psi_0(x_2)\right]
\end{aligned}\right\} \quad \text{with energy } E = 2\hbar\omega.$$

The possible total spin states are singlet and triplet. Denoting by χ_{s,s_z} the total spin $\mathbf{S}$ eigenstates with eigenvalues $S^2 = s(s+1)\hbar^2$ and $S_z = s_z\hbar$, we have an antisymmetric state and three symmetric states:

$$\begin{aligned}
\chi_{0,0} &= \frac{1}{\sqrt{2}}\left[\chi_+(1)\,\chi_-(2) - \chi_-(1)\,\chi_+(2)\right], \\
\chi_{1,-1} &= \chi_-(1)\,\chi_-(2), \\
\chi_{1,0} &= \frac{1}{\sqrt{2}}\left[\chi_+(1)\,\chi_-(2) + \chi_-(1)\,\chi_+(2)\right], \\
\chi_{1,+1} &= \chi_+(1)\,\chi_+(2).
\end{aligned}$$

The fully antisymmetric overall wave functions are

$$\psi_0^+(x_1, x_2)\,\chi_{0,0} \qquad \text{for the ground state}$$

and

$$\left.\begin{array}{l} \psi_1^+ (x_1, x_2) \, \chi_{0,0} \\ \psi_1^- (x_1, x_2) \, \chi_{1,-1} \\ \psi_1^- (x_1, x_2) \, \chi_{1,0} \\ \psi_1^- (x_1, x_2) \, \chi_{1,+1} \end{array}\right\} \quad \text{for the first excited state.}$$

8.5 Double Oscillator for Identical Particles

Two particles, each of mass m, are confined in the potential of a one-dimensional harmonic oscillator

$$V = \frac{1}{2} k x^2$$

and interact with each other through another attractive elastic force having an elastic constant $\bar{k} \ll k$.

(a) What are the states corresponding to the three lowest levels of the system Hamiltonian?

(b) If the particles are identical and have 0 spin, which of these three states are allowed?

(c) If the particles are identical and have spin $\frac{1}{2}$, what is the total spin of each of these three states?

(Hint: use an appropriate change of variables.)

Solution

Having denoted the coordinates of the two particles by x_1 e x_2, the system's Hamiltonian is

$$\mathcal{H} = -\frac{\hbar^2}{2m} \left(\frac{\partial^2}{\partial x_1^2} + \frac{\partial^2}{\partial x_2^2} \right) + \frac{1}{2} k (x_1^2 + x_2^2) + \frac{1}{2} \bar{k} (x_1 - x_2)^2.$$

We introduce the relative and center of mass coordinates

$$X = \frac{x_1 + x_2}{2} \quad \text{and} \quad x = x_1 - x_2.$$

In these variables, the Hamiltonian itself appears as a sum of terms relative to harmonic oscillators, each one dependent on a single variable:

$$\mathcal{H} = \mathcal{H}_{CM} + \mathcal{H}_r,$$
$$\mathcal{H}_{CM} = -\frac{\hbar^2}{2M} \frac{\partial^2}{\partial X^2} + \frac{1}{2} M \omega^2 X^2,$$
$$\mathcal{H}_r = -\frac{\hbar^2}{2\mu} \frac{\partial^2}{\partial x^2} + \frac{1}{2} \mu \omega'^2 x^2,$$

where we introduced the total mass of the system and the reduced mass

$$M = 2m \quad \text{and} \quad \mu = \frac{m}{2}$$

and the frequencies

$$\omega = \sqrt{\frac{k}{m}} \quad \text{and} \quad \omega' = \sqrt{\frac{k + 2\bar{k}}{2\mu}}.$$

The separation of variables allows us to solve the eigenvalue equations for the two Hamiltonians separately: the eigenfunctions are the product of the eigenfunctions of the individual oscillators and the eigenvalues are the sum of the eigenvalues.

(a) To determine the lowest energy levels, the hypothesis $\bar{k} \ll k$ must be taken into account, from which we derive $2\hbar\omega > \hbar\omega' > \hbar\omega$. So, the three lowest levels of energy are, in ascending order,

$$E_0 = \frac{1}{2}\hbar(\omega + \omega'),$$
$$E_1 = E_0 + \hbar\omega,$$
$$E_2 = E_0 + \hbar\omega'.$$

Having denoted by $\phi_n^{(\omega)}(x)$ the nth eigenfunction of a harmonic oscillator with frequency ω, the three corresponding eigenfunctions are given by

$$\psi_0(x_1, x_2) = \phi_0^{(\omega)}(X)\,\phi_0^{(\omega')}(x),$$
$$\psi_1(x_1, x_2) = \phi_1^{(\omega)}(X)\,\phi_0^{(\omega')}(x),$$
$$\psi_2(x_1, x_2) = \phi_0^{(\omega)}(X)\,\phi_1^{(\omega')}(x).$$

(b) If the particles are identical bosons, the wave function must be symmetric for the exchange of coordinates $x_1 \leftrightarrow x_2$, corresponding to the exchange $x \leftrightarrow -x$. Therefore, taking into account the fact that the nth eigenfunction of the harmonic oscillator energy has parity $(-1)^n$, the third level is not allowed.

(c) If the particles are fermions, the spatial wave functions must be multiplied by the total spin eigenfunction, which is antisymmetric in the case of spin 0 (singlet) and symmetric in the case of spin 1 (triplet). If the fermions are identical, the wave function must be overall antisymmetric by exchange. We will therefore have the following total spin states for each of the 3 levels:

$$E_0 \rightarrow S = 0,$$
$$E_1 \rightarrow S = 0,$$
$$E_2 \rightarrow S = 1.$$

8.6 Identical Particles in a Box

Two non-interacting identical particles of mass m are closed in the box $|x| < a, |y| < b, |z| < c$, with $a > b > c > 0$.

(a) Determine the energy eigenvalues and eigenfunctions, specifying the degree of degeneracy for the ground level and the first excited level, in the case of spin-free particles and in the case of fermions.
(b) Assuming the particles are fermions in one of the eigenstates common to the total spin operators (S^2 and S_z) and the Hamiltonian in the first excited state, determine the probability of finding both particles in the $x > 0$ region.
(c) Suppose, still in the case of fermions, that we add to the Hamiltonian the term

$$\lambda V = \lambda \frac{\pi^2 \hbar^2}{8ma^4} \mathbf{r_1} \cdot \mathbf{r_2} \qquad (\lambda << 1).$$

Determine the first-order perturbative corrections in the first two levels.

Solution

The Hamiltonian is separable into the three coordinates, each of which is bound on a segment of different length. Each of the particles, if present individually, would have energy eigenfunctions

$$\psi_{k,l,m}(\mathbf{r}) = \psi_k(x)\, \psi_l(y)\, \psi_m(z),$$

where

$$\psi_k(x) = \begin{cases} \sqrt{\frac{1}{a}}\, \cos \frac{k\pi x}{2a}, & \text{odd } k \\ \sqrt{\frac{1}{a}}\, \sin \frac{k\pi x}{2a}, & \text{even } k \end{cases}$$

within the $(-a, a)$ segment and zero outside. Similar expressions are valid for $\psi_l(y)$ and $\psi_m(z)$. The corresponding eigenvalues would be

$$E_{k,l,m} = \frac{\pi^2 \hbar^2}{8m} \left(\frac{k^2}{a^2} + \frac{l^2}{b^2} + \frac{m^2}{c^2} \right).$$

(a) In the absence of spin, and therefore of symmetry by exchange, the ground state is non-degenerate, has energy equal to $2 E_{1,1,1}$ and wave function

$$\psi_{1,1,1}(\mathbf{r_1})\, \psi_{1,1,1}(\mathbf{r_2}),$$

which is symmetric for exchange of the two particles.
The first excited level is obtained by increasing the quantum number relative to the coordinate x to $k = 2$ (remember that $a > b > c$). It is doubly degenerate, has energy $E_{1,1,1} + E_{2,1,1}$ and eigenfunctions

$$\psi_{1,1,1}(\mathbf{r_1})\,\psi_{2,1,1}(\mathbf{r_2})$$

and

$$\psi_{2,1,1}(\mathbf{r_1})\,\psi_{1,1,1}(\mathbf{r_2}),$$

which can be combined symmetrically or antisymmetrically.

In the case of fermions, it is necessary to impose the antisymmetry of the complete wave function (position and spin) for particle exchange. The energy levels do not change, while, as regards the self-functions, the situation is similar to that of problem Sect. 8.4. Using the same notation for spin states, the ground state is non-degenerate and has wave function

$$\psi_{1,1,1}(\mathbf{r_1})\,\psi_{1,1,1}(\mathbf{r_2})\,\chi_{0,0},$$

while the first excited level is four times degenerate, has energy equal to $E_{1,1,1} + E_{2,1,1}$ and wave functions

$$\psi_1^+(\mathbf{r_1},\mathbf{r_2})\,\chi_{0,0},$$
$$\psi_1^-(\mathbf{r_1},\mathbf{r_2})\,\chi_{1,-1},$$
$$\psi_1^-(\mathbf{r_1},\mathbf{r_2})\,\chi_{1,0},$$
$$\psi_1^-(\mathbf{r_1},\mathbf{r_2})\,\chi_{1,+1},$$

where

$$\psi_1^+(\mathbf{r_1},\mathbf{r_2}) = \frac{1}{\sqrt{2}}\left[\psi_{1,1,1}(\mathbf{r_1})\,\psi_{2,1,1}(\mathbf{r_2}) + \psi_{2,1,1}(\mathbf{r_1})\,\psi_{1,1,1}(\mathbf{r_2})\right] =$$

$$= \frac{1}{\sqrt{2}}\left[\psi_1(x_1)\,\psi_2(x_2) + \psi_2(x_1)\,\psi_1(x_2)\right]\,\psi_1(y_1)\,\psi_1(z_1)\,\psi_1(y_2)\,\psi_1(z_2),$$

$$\psi_1^-(\mathbf{r_1},\mathbf{r_2}) = \frac{1}{\sqrt{2}}\left[\psi_{1,1,1}(\mathbf{r_1})\,\psi_{2,1,1}(\mathbf{r_2}) - \psi_{2,1,1}(\mathbf{r_1})\,\psi_{1,1,1}(\mathbf{r_2})\right] =$$

$$= \frac{1}{\sqrt{2}}\left[\psi_1(x_1)\,\psi_2(x_2) - \psi_2(x_1)\,\psi_1(x_2)\right]\,\psi_1(y_1)\,\psi_1(z_1)\,\psi_1(y_2)\,\psi_1(z_2).$$

(b) After integrating over coordinates y and z, we calculate the desired probability:

$$P_\pm(x_1, x_2 > 0) = \frac{1}{2}\int_0^a dx_1 \int_0^a dx_2\,[\psi_1(x_1)\,\psi_2(x_2) \pm \psi_2(x_1)\,\psi_1(x_2)]^2 =$$

$$= \frac{1}{2}\int_0^a dx_1 \int_0^a dx_2\,[\psi_1^2(x_1)\,\psi_2^2(x_2) + \psi_2^2(x_1)\,\psi_1^2(x_2) \pm$$
$$\pm 2\psi_1(x_1)\,\psi_2(x_2)\psi_2(x_1)\,\psi_1(x_2)] =$$

$$= \frac{1}{2}\left\{\frac{1}{2} \pm 2\left[\int_0^a dx\,\psi_1(x)\,\psi_2(x)\right]^2\right\} =$$

$$= \frac{1}{2}\left\{\frac{1}{2} \pm \frac{2}{a^2}\left[\int_0^a dx\,\cos\frac{\pi x}{2a}\,\sin\frac{\pi x}{a}\right]^2\right\} =$$

$$= \frac{1}{2} \left\{ \frac{1}{2} \pm 2 \left[\frac{4}{3\pi} \right]^2 \right\}.$$

(c) For each state $|\psi\rangle$, we have to calculate

$$\langle \psi | \lambda V | \psi \rangle = \lambda \frac{\pi^2 \hbar^2}{8ma^4} I,$$

where I is

$$I = \langle \psi | x_1 x_2 + y_1 y_2 + z_1 z_2 | \psi \rangle.$$

For the ground state we have

$$I = \langle x_1 \rangle \langle x_2 \rangle + \langle y_1 \rangle \langle y_2 \rangle + \langle z_1 \rangle \langle z_2 \rangle = 0,$$

because, for reasons of symmetry, each expectation value is zero.
For the first excited level, we have to diagonalize the matrix relative to the two different wave functions (the potential does not depend on the spin). Let us first consider the diagonal elements, for which we immediately see that the coordinates y and z do not contribute. For the first diagonal term, we have

$$I_{+,+} = \langle \psi_1^+ | x_1 x_2 | \psi_1^+ \rangle$$
$$= \frac{1}{2} \int_{-a}^{a} dx_1 \, dx_2 \, x_1 x_2 \left[\psi_1(x_1) \, \psi_2(x_2) + \psi_2(x_1) \, \psi_1(x_2) \right]^2.$$

Once the square has been expanded, we note that the two quadratic terms give zero result, because they are proportional to the expectation value of x in an eigenstate of the well. We obtain, therefore, integrating by parts,

$$I_{+,+} = \left[\int_{-a}^{a} dx \, x \, \psi_1(x) \, \psi_2(x) \right]^2 = 2 \left(\frac{32a}{9\pi^2} \right)^2.$$

In a similar way, we obtain

$$I_{-,-} = \langle \psi_1^- | x_1 x_2 | \psi_1^- \rangle = -2 \left(\frac{32a}{9\pi^2} \right)^2.$$

It is easy to see, with analogous calculations, that the non-diagonal terms are null. So, the corrections in the first excited level are

$$\langle \psi_1^\pm | \lambda V | \psi_1^\pm \rangle = \pm \frac{128}{81} \lambda \frac{\hbar^2}{ma^2\pi^2}.$$

8.7 Three Interacting Fermions on a Segment

Three identical particles of mass m and spin $1/2$ are constrained to move along the x-axis in the segment $(0, a)$. They are subject to potential

$$V = \alpha \left(\mathbf{S}_1 \cdot \mathbf{S}_2 + \mathbf{S}_1 \cdot \mathbf{S}_2 + \mathbf{S}_2 \cdot \mathbf{S}_3 \right),$$

with $\alpha < 0$ and $\mathbf{S}_i$ spin of the ith particle. A measure of the total spin z-component provides the value $S_z = +\frac{3}{2}\hbar$. Write the possible energy eigenfunctions and the related eigenvalues. What is the eigenfunction of the ground state of the system in this spin state? What is the energy eigenvalue in this state?

Solution

The potential can be rewritten in the form

$$V = \frac{\alpha}{2} \left(S^2 - S_1^2 - S_2^2 - S_3^2 \right),$$

therefore, the energy eigenfunctions are the product of eigenfunctions related to the spatial coordinates for total spin eigenfunctions. The three fermions are in a state with $S_z = +\frac{3}{2}\hbar$ and spin quantum number $s = \frac{3}{2}$, which is the maximum value that s can assume. The spins of the three particles have the same component z, so the total spin eigenfunction is symmetric and, as a consequence, the part dependent on the spatial coordinates must be antisymmetric:

$$\Psi_{n_1,n_2,n_3; \frac{3}{2},+\frac{3}{2}}(x_1, x_2, x_3) = \frac{1}{\sqrt{3!}} \det \begin{vmatrix} \psi_{n_1}(x_1) & \psi_{n_1}(x_2) & \psi_{n_1}(x_3) \\ \psi_{n_2}(x_1) & \psi_{n_2}(x_2) & \psi_{n_2}(x_3) \\ \psi_{n_3}(x_1) & \psi_{n_3}(x_2) & \psi_{n_3}(x_3) \end{vmatrix} \chi_{\frac{3}{2},\frac{3}{2}}.$$

The eigenvalues corresponding to these eigenfunctions are

$$E = \frac{\pi^2 \hbar^2}{2ma^2} (n_1^2 + n_2^2 + n_3^2) + \frac{3}{4} \alpha \hbar^2 \quad \text{with } n_1, n_2, n_3 = 1, 2, \ldots$$

Given the spin state, the spatial part of the eigenfunction must also be antisymmetric in the ground state, so the three quantum numbers must be different and such that the energy is minimal, thus $n_1 = 1$, $n_2 = 2$, $n_3 = 3$. So, we get

$$\Psi_{1,2,3; \frac{3}{2},+\frac{3}{2}}(x_1, x_2, x_3).$$

The corresponding eigenvalue is

$$E = \left(\frac{7\pi^2}{ma^2} + \frac{3}{4}\alpha \right) \hbar^2.$$

8.8   Two Interacting Fermions in a Sphere

Two identical particles of spin $\frac{1}{2}$ are in an impenetrable sphere of radius $R.$.

(a) Write the energy eigenfunctions concerning the ground state and the first excited state.
(b) Calculate the way in which the degeneracy is modified if the following interaction occurs:

$$V = \alpha \mathbf{S}_1 \cdot \mathbf{S}_2,$$

where $\mathbf{S}_1$ and $\mathbf{S}_2$ are the two particles' spins.

Solution

In the case of a single spin $\frac{1}{2}$ particle in an impenetrable sphere, having in mind the results of Problem 3.13, the eigenfunctions common to energy and spin are given by

$$\psi_{n,\ell,m,m_s}(r, \theta, \varphi) = \psi_{n,\ell,m}(r, \theta, \varphi) \, \chi_{m_s},$$

where χ_{m_s} are the S_z eigenstates corresponding to eigenvalues $m_s \hbar = \pm \frac{\hbar}{2}$ and

$$\psi_{n,\ell,m}(r, \theta, \varphi) = R_{n,\ell}(r) \, Y_\ell^m(\theta, \varphi),$$

where

$$R_{n,\ell}(r) = N \, j_\ell(k_{n,\ell} \, r) \quad \text{and} \quad k_{n,\ell} = \frac{z_{n,\ell}}{R}.$$

Here, j_ℓ is the ℓth spherical Bessel function, while $z_{n,\ell}$ is the nth zero of j_ℓ. $k_{n,\ell}$ is related to the energy eigenvalue $E_{n,\ell}$ through the relation

$$E_{n,\ell} = \frac{\hbar^2}{2m} k_{n,\ell}^2.$$

The ground state is not degenerate and is obtained for $\ell = 0$ and $n = 1$,

$$k_{1,0} = \frac{\pi}{R}, \qquad E_{1,0} = \frac{\hbar^2 \pi^2}{2m R^2}.$$

The first excited state is obtained for $\ell = 1$ and $n = 1$ and is three times degenerate ($m = 0, \pm 1$).

Now we come to the case of two spin $\frac{1}{2}$ particles.

(a) The two fermions' wave function must be completely antisymmetric for particle exchange. In the case of the ground state, there is only one possibility, corresponding to the singlet state for the total spin, $\chi_{S^2, S_z} = \chi_{0,0}$. So, we have

$$\psi_{1,s=0}(\mathbf{r}_1, \mathbf{r}_2) = \psi_{1,0,0}(\mathbf{r}_1), \psi_{1,0,0}(\mathbf{r}_2) \, \chi_{0,0}.$$

The first excited level of the system must have one particle in the ground state and the other in the first excited state. This allows both the singlet state

$$\psi_{2,m,s=0}(\mathbf{r}_1, \mathbf{r}_2) = \frac{1}{\sqrt{2}} \left[\psi_{1,0,0}(\mathbf{r}_1), \psi_{1,1,m}(\mathbf{r}_2) + \psi_{1,1,m}(\mathbf{r}_1), \psi_{1,0,0}(\mathbf{r}_2) \right] \chi_{0,0},$$

which has 3 possible determinations in correspondence of the 3 values of $m = 0, \pm 1$, and the triplet state

$$\psi_{2,m,s=1}(\mathbf{r}_1, \mathbf{r}_2) = \frac{1}{\sqrt{2}} \left[\psi_{1,0,0}(\mathbf{r}_1), \psi_{1,1,m}(\mathbf{r}_2) - \psi_{1,1,m}(\mathbf{r}_1), \psi_{1,0,0}(\mathbf{r}_2) \right] \chi_{1,m_s},$$

which has 9 possible determinations at the 3 values of $m = 0, \pm 1$ and the 3 values of $m_s = 0, \pm 1$.

(b) The interaction $V = \alpha \mathbf{S}_1 \cdot \mathbf{S}_2$ can be rewritten in the form

$$V = \frac{\alpha}{2} (S^2 - S_1^2 - S_2^2) = \frac{\alpha}{2} (S^2 - \frac{3}{2} \hbar^2).$$

In the singlet states, $V = -\frac{3}{4} \alpha \hbar^2$, while, in the triplet states, $V = \frac{1}{4} \alpha \hbar^2$. So, there is a different contribution to the energy eigenvalue, depending on the total spin state.

With regard to degeneracy, it is still absent in the case of the ground state with a change in the energy value. Degeneracy is only present in the case of the first excited state, which now gives rise to two distinct levels:

$$E_{1,s=0} = \frac{\hbar^2}{2m R^2} (z_{1,1}^2 + \pi^2) - \frac{3}{4} \alpha \hbar^2 \quad \text{3 times degenerate,}$$

$$E_{1,s=1} = \frac{\hbar^2}{2m R^2} (z_{1,1}^2 + \pi^2) + \frac{1}{4} \alpha \hbar^2 \quad \text{9 times degenerate,}$$

where $z_{1,1} = 4.49341$ (see [1]).

8.9 Two Fermions on the Surface of a Sphere

Two identical particles of spin $\frac{1}{2}$, not interacting with each other, are constrained to move along a spherical surface, so that the only degrees of freedom are angular position and spin. Among all of the states accessible to the two-particle system, determine how many states meet the following properties:

(a) are eigenstates of both L_z and S_z, the components along the z axis of total angular momentum $\mathbf{L} = \mathbf{L}_1 + \mathbf{L}_2$ and of the total spin $\mathbf{S} = \mathbf{S}_1 + \mathbf{S}_2$, where indices 1 and 2 refer to the two particles;

(b) the quantum numbers ℓ_1 and ℓ_2 relative to the angular momenta of the two particles are either 0 or 1;

(c) the total spin state is a triplet one (eigenstate of S^2 with $s = 1$).

Solution

The desired states are the product of the eigenstates of the total angular momentum times the total spin eigenstates.

The third condition indicates that the eigenstate of the total angular momentum must be antisymmetric for the exchange of the two particles. Since, for the second condition, $\ell_1 = 0, 1$ and $\ell_2 = 0, 1$, we have the following possibilities:

1. $\ell_1 = 0$ and $\ell_2 = 0$: there is only one state, $|m_1, m_2\rangle = |0, 0\rangle$, which is symmetric, and therefore to be excluded;

2. $\ell_1 = 0$ and $\ell_2 = 1$: there are three states, $|0, 0\rangle$, $|0, +1\rangle$, $|0, -1\rangle$,

3. $\ell_1 = 1$ and $\ell_2 = 0$: there are three states, $|0, 0\rangle$, $|+1, 0\rangle$, $|-1, 0\rangle$; both in this case and in the previous one, the states have no definite symmetry, but, starting from these six states, we can consider the three symmetric linear combinations and the three antisymmetric ones;

4. $\ell_1 = 1$ and $\ell_2 = 1$: $\ell = 0, 1, 2$ total angular momentum states are possible. Since Clebsch-Gordan coefficients meet the following relationship:

$$\langle j_1, j_2, m_1, m_2 | j_1, j_2, J, M \rangle =$$
$$(-1)^{J-j_1-j_2} \quad \langle j_2, j_1, m_2, m_1 | j_2, j_1, J, M \rangle,$$

it turns out that states with $\ell = 0, 2$ are symmetric and should not be considered. Instead, the states corresponding to $\ell = 1$ are antisymmetric:

$$|\ell = 1, m = 0\rangle = \frac{1}{\sqrt{2}} \left(|m_1 = +1, m_2 = -1\rangle - |m_1 = -1, m_2 = +1\rangle \right),$$

$$|\ell = 1, m = +1\rangle = \frac{1}{\sqrt{2}} \left(|m_1 = +1, m_2 = 0\rangle - |m_1 = 0, m_2 = +1\rangle \right),$$

$$|\ell = 1, m = -1\rangle = \frac{1}{\sqrt{2}} \left(|m_1 = -1, m_2 = 0\rangle - |m_1 = 0, m_2 = -1\rangle \right).$$

The answer, therefore, is that there are a total of 6 states with respect to the angular momentum multiplied by 3, the possible total spin states, giving the total number of 18 states.

8.10 Three Electrons in a Central Potential

Three electrons are bound by a central potential, and the interactions between the electrons are to be ignored. One of them is in the energy eigenstate E_1 with spatial wave function ψ_1, while the other two are in the energy eigenstate E_2 with spatial wave function ψ_2.

(a) Write the possible states that are compatible with the Fermi-Dirac statistic and the value of their spin in these states.

(b) What is the total degeneracy of the level $E = E_1 + 2E_2$ if ψ_1 and ψ_2 both correspond to states with $\ell = 0$?

Solution

(a) Denote, by $\psi_n^\pm$, the eigenfunction of an electron that is located in the energy level n and in the state of spin with $m_s = \pm 1/2$. The two particles in the state with $n = 2$ must be in different spin states, while the other particle can be indifferently in one of the two spin states. Altogether, we have two possibilities that we can write, antisymetrizing through the determinant of Slater,

$$\Psi_{1,2,2;\, j=\frac{1}{2},m_j=+\frac{1}{2}}(x_1, x_2, x_3) = \frac{1}{\sqrt{3!}} \det \begin{vmatrix} \psi_1^+(x_1) & \psi_1^+(x_2) & \psi_1^+(x_3) \\ \psi_2^+(x_1) & \psi_2^+(x_2) & \psi_2^+(x_3) \\ \psi_2^-(x_1) & \psi_2^-(x_2) & \psi_2^-(x_3) \end{vmatrix}$$

and

$$\Psi_{1,2,2;\, j=\frac{1}{2},m_j=-\frac{1}{2}}(x_1, x_2, x_3) = \frac{1}{\sqrt{3!}} \det \begin{vmatrix} \psi_1^-(x_1) & \psi_1^-(x_2) & \psi_1^-(x_3) \\ \psi_2^+(x_1) & \psi_2^+(x_2) & \psi_2^+(x_3) \\ \psi_2^-(x_1) & \psi_2^-(x_2) & \psi_2^-(x_3) \end{vmatrix}.$$

In the previous expressions, j and m_j are the quantum numbers related to the total spin. In fact, it is easy to see, expanding the determinants according to the first line, that the two electrons in the level with $n = 2$ are in a singlet state; therefore, their total spin is 0 and the total spin of the three electrons must be $1/2$. The value of its z component is then given by the spin status of the electron in the level with $n = 1$.

(b) Since there is no degeneracy due to angular momentum ($\ell = 0$), the degeneracy of the $E = E_1 + 2E_2$ level is 2.

Chapter 9
Scattering (Born Approximation)

9.1 Yukawa and Coulomb Potential

Determine the Born approximation of the differential cross-section for the elastic scattering from the:

(a) Yukawa potential:

$$V(r) = V_0 \frac{e^{-\alpha r}}{\alpha r};$$

(b) Coulomb potential:

$$V(r) = \frac{q_1 q_2}{r}.$$

Solution

The scattering amplitude in the Born approximation is given by (A.84).

(a) For the Yukawa potential, we get

$$
\begin{aligned}
f_B(q) &= -\frac{2m V_0}{\hbar^2 q} \int_0^\infty dr \, \sin(qr) \frac{e^{-\alpha r}}{\alpha r} \, r = \\
&= -\frac{2m V_0}{\hbar^2 q \alpha} \Im \left(\int_0^\infty dr \, e^{-\alpha r + \iota q r} \right) = \\
&= -\frac{2m V_0}{\hbar^2 q \alpha} \Im \left(\frac{1}{\alpha - \iota q} \right) = \\
&= -\frac{2m V_0}{\hbar^2 \alpha} \frac{1}{\alpha^2 + q^2}.
\end{aligned}
$$

The differential cross-section is

$$\frac{d\sigma_B}{d\Omega} = \left(\frac{2m V_0}{\hbar^2 \alpha} \right)^2 \left(\frac{1}{\alpha^2 + 4k^2 \sin^2 \frac{\theta}{2}} \right)^2,$$

© Springer Nature Switzerland AG 2019
L. Angelini, *Solved Problems in Quantum Mechanics*, UNITEXT for Physics,
https://doi.org/10.1007/978-3-030-18404-9_9

where θ is the angle between the incident and the scattering direction.

(b) We can obtain the differential cross-section for the Coulomb potential from the previous formulas through the limit

$$\alpha \to 0, \qquad V_0 \to 0, \qquad \frac{V_0}{\alpha} \to q_1 q_2.$$

The result is

$$\frac{d\sigma_B}{d\Omega} = \frac{q_1^2 q_2^2}{16 E^2 \sin^4 \frac{\theta}{2}},$$

where $E = \hbar^2 k^2 / 2m$ is the energy of the particle incident on the center of force. This result coincides with the classic Rutherford one and with the exact quantum result (notice that the cross-section does not depend on $\hbar$).

9.2   Gaussian Potential

Calculate, in the Born approximation, the differential and total cross-sections for elastic scattering from the potential

$$V(x) = V_0 e^{-\alpha^2 r^2}.$$

Solution

Using the notation of Appendix A.13, we apply formula (A.84)

$$
\begin{aligned}
f_B(q) &= -\frac{2m}{\hbar^2 q} \int_0^\infty dr \, \sin(qr) \, V(r) \, r = \\
&= -\frac{2m V_0}{\hbar^2 q} \int_0^\infty dr \, \sin(qr) \, r \, e^{-\alpha^2 r^2} = \\
&= -\frac{\sqrt{\pi} m V_0}{2\hbar^2 \alpha^3} e^{-\frac{q^2}{4\alpha^2}}.
\end{aligned}
$$

In the last step, we used formula (A.5). Considering $q = 2k \sin \frac{\theta}{2}$, the differential cross-section in this approximation is

$$\frac{d\sigma}{d\Omega} = |f_B(q)|^2 = \frac{\pi \mu^2 V_0^2}{4\hbar^4 \alpha^6} e^{-2\frac{k^2}{\alpha^2} \sin^2 \frac{\theta}{2}}.$$

Finally, we calculate the total cross-section:

$$\sigma = \int d\Omega \frac{d\sigma}{d\Omega} = 2\pi \frac{\pi m^2 V_0^2}{4\hbar^4 \alpha^6} \int_{-1}^{+1} d\cos\theta\, e^{-\frac{k^2}{\alpha^2}(1-\cos\theta)} =$$

$$= \frac{\pi^2 m^2 V_0^2}{2\hbar^4 \alpha^6} \left(1 - e^{-2\frac{k^2}{\alpha^2}}\right) =$$

$$= \frac{\pi^2 m V_0^2}{4\hbar^2 \alpha^6 E} \left(1 - e^{-4m\frac{E}{\hbar^2 \alpha^2}}\right),$$

where $E = \frac{\hbar^2 k^2}{2m}$ is the energy of the scattered particles.

9.3   Scattering From an Opaque Sphere

Determine the Born approximation of the differential and total cross-sections for elastic scattering from the potential

$$V(r) = \begin{cases} -V_0 & \text{if } r \in [0, a], \\ 0 & \text{elsewhere.} \end{cases}$$

Solution

Using the notation of Appendix A.13, we apply formula (A.84)

$$f_B(q) = -\frac{2m}{\hbar^2 q} \int_0^\infty dr\, \sin(qr)\, V(r)\, r = \frac{2m V_0}{\hbar^2 q} \int_0^a dr\, \sin(qr)\, r =$$

$$= \frac{2m V_0}{\hbar^2 q^3} \int_0^{qa} dx\, \sin x\, x = \frac{2m V_0}{\hbar^2 q^3} \left[\sin(qa) - qa\cos(qa)\right] =$$

$$= \frac{2m V_0 a^3}{\hbar^2} \varphi(qa),$$

where

$$\varphi(x) = \frac{\sin x - x\cos x}{x^3} = \frac{1}{x} j_1(x).$$

and j_p are the spherical Bessel functions (see (A.47)). The differential cross-section is therefore

$$\frac{d\sigma_B}{d\Omega} = \left(\frac{2m V_0 a^3}{\hbar^2}\right)^2 \varphi^2(2ka\sin\frac{\theta}{2}),$$

where θ is the angle of deflection.

Let us study the behavior of the differential cross-section. Since

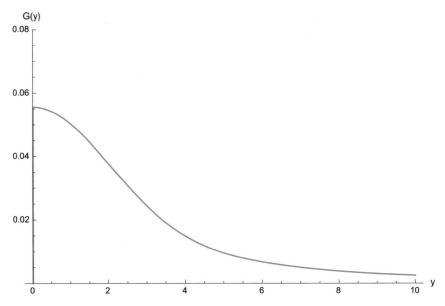

Fig. 9.1 Scattering from an opaque sphere: behavior in the variable $y = 2ka$ of the Born total cross-section (apart from an overall constant)

$$\lim_{x \to 0} \varphi(x) = \frac{1}{3},$$
$$\lim_{x \to \infty} \varphi(x) = 0,$$

the differential cross-section shows a maximum in $\theta = 0$, k^{-2} - damped oscillations and zeros placed where $x = \tan x$, with $x = 2ka \sin \frac{\theta}{2}$. For $ka \ll 1$, the scattering is isotropic; for $ka \gg 1$, as the cross-section goes quickly to zero, the diffusion takes place in an essentially forward manner.

We now calculate the total cross-section in Born approximation.

$$\sigma_B = \int d\Omega \frac{d\sigma_B}{d\Omega} =$$

$$= 2\pi \left(\frac{2m V_0 a^3}{\hbar^2} \right)^2 \int_0^\pi d\theta\, 2 \sin \frac{\theta}{2} \cos \frac{\theta}{2} \varphi^2(2ka \sin \frac{\theta}{2}) =$$

$$= 8\pi \left(\frac{2m V_0 a^3}{\hbar^2} \right)^2 \frac{1}{(2ka)^2} \int_0^{2ka} dx\, x\, \varphi^2(x) =$$

$$= 8\pi \left(\frac{2m V_0 a^3}{\hbar^2} \right)^2 G(2ka),$$

where

$$G(y) = \frac{1}{y^2} \int_0^y dx \, \frac{(\sin x - x \cos x)^2}{x^5} = \frac{\cos 2y + 2y \sin 2y + 2y^4 - 2y^2 - 1}{8y^6}.$$

This function tends to a constant value of $1/18$ for $y \to 0$ and is monotone decreasing. Its behavior is shown in Fig. 9.1

Chapter 10
WKB Approximation

10.1 Energy Spectrum of the Harmonic Oscillator

Calculate the energy spectrum of the harmonic oscillator in WKB approximation.

Solution

The WKB method calculates an approximation to energy eigenvalue E_n energy imposing that relation (A.86) is satisfied, i.e., that the integral of the impulse along the classical closed trajectory (one period) for that energy is equal to $n + \frac{1}{2}$ times $2\pi\hbar$, where n is an integer. This integral is equal to the area included within the classical trajectory in the $p - q$ phase space. For each energy level E_n, such a trajectory is defined by the relationship

$$E_n = \frac{p^2}{2m} + \frac{1}{2}m\omega^2 q^2,$$

i.e.,

$$1 = \frac{p^2}{2mE_n} + \frac{q^2}{\frac{2E_n}{m\omega^2}}.$$

In the $p - q$ phase space, the closed path corresponding to a period of classical motion is, therefore, an ellipsis of semi-axes

$$a = \sqrt{\frac{2E_n}{m\omega^2}} \quad \text{and} \quad b = \sqrt{2mE_n},$$

whose area is

$$\pi ab = \pi \frac{2E_n}{\omega}.$$

Applying the WKB quantization condition (A.86), we obtain

© Springer Nature Switzerland AG 2019
L. Angelini, *Solved Problems in Quantum Mechanics*, UNITEXT for Physics,
https://doi.org/10.1007/978-3-030-18404-9_10

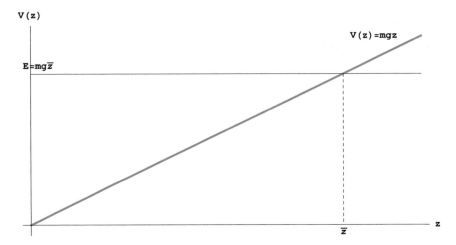

Fig. 10.1 Potential well for a body falling to the Earth's surface

$$\pi \frac{2E_n}{\omega} = 2\pi\hbar\left(n + \frac{1}{2}\right) \qquad \text{with } n = 0, 1, 2\ldots,$$

i.e.,

$$E_n = \left(n + \frac{1}{2}\right)\hbar\omega \qquad \text{with } n = 0, 1, 2\ldots.$$

In this case, the WKB method gives the exact result.

10.2   Free Fall of a Body

Determine the energy spectrum for a body falling to the Earth's surface in WKB approximation. Show that, for macroscopic objects, it is not possible to reveal quantum effects.

Solution

The free fall of a body to the Earth's surface is the motion of a mass m in a potential

$$V(z) = \begin{cases} mgz, & \text{if } x > 0 \\ \infty, & \text{if } x < 0 \end{cases},$$

shown in Fig. 10.1. Classically, supposing elastic bumps against the floor at $z = 0$, there is a periodic motion between the turning points $z = 0$ and $z = \bar{z} = \frac{E}{mg}$. The period T can be obtained through the relationship

$$\frac{1}{2} g \left(\frac{T}{2}\right)^2 = \frac{E}{mg},$$

and the corresponding frequency is

$$\omega = \pi g \sqrt{\frac{m}{2E}}.$$

The discrete energy spectrum can be evaluated in WKB approximation through the Bohr-Sommerfeld relation:

$$I = \int_0^{\bar{z}} dz \sqrt{2m(E_n - mgz)} = \pi \hbar \left(n + \frac{1}{2}\right).$$

We notice, however, that this rule was derived under the hypothesis that the potential is weakly variable, and this certainly does not happen in $z = 0$, where $V(z)$ diverges abruptly. However, we can resort to the trick of considering the potential extension to $z < 0$ in the form

$$\bar{V}(z) = k|z|, \tag{10.1}$$

restricting ourselves, however, to odd solutions (odd n) that are zero in the origin. We must therefore impose

$$\int_{-\bar{z}}^{\bar{z}} dz \sqrt{2m(E_n - mg|z|)} = \pi \hbar \left(n + \frac{1}{2}\right) \qquad \text{with } n = 1, 3, 5, \ldots$$

or, equivalently, given the potential symmetry,

$$I = \int_0^{\bar{z}} dz \sqrt{2m(E_n - mgz)} = \pi \hbar \left(n - \frac{1}{4}\right) \qquad \text{with } n = 1, 2, 3, \ldots.$$

This integral can be easily evaluated:

$$I = \sqrt{2m^2 g} \int_0^{\bar{z}} dz \sqrt{\bar{z} - z} = -\sqrt{2m^2 g} \left[\frac{2}{3} (\bar{z} - z)^{\frac{3}{2}}\right]_0^{\bar{z}} = \frac{2}{3} \sqrt{2m^2 g} \, \bar{z}^{\frac{3}{2}}.$$

We therefore obtain the spectrum

$$E_n = \frac{\left[3\pi \left(n - \frac{1}{4}\right)\right]^{\frac{2}{3}}}{2} \sqrt[3]{mg^2 \hbar^2} \qquad \text{with } n = 1, 2, 3, \ldots.$$

At a macroscopic level, quantum effects are not perceptible. Indeed, in the case of a mass $m = 1 \, Kg$ that falls from one meter in height, the energy is

$$E_n = mg\bar{z} = 9.8 J,$$

while the ground state energy is

$$E_1 \simeq \sqrt[3]{mg^2\hbar^2} \simeq (1 \cdot 10^2 \cdot 10^{-68})^{\frac{1}{3}} = 10^{-22} \ J.$$

Let us estimate n. Since, for large n, it results that

$$E_n \sim n^{\frac{2}{3}} E_1,$$

we have

$$n \sim \left(\frac{E_n}{E_1}\right)^{\frac{3}{2}} \simeq \left(\frac{9.8}{10^{-22}}\right)^{\frac{3}{2}} \simeq 10^{34}.$$

This is a huge number, but it may be that, around the E_n level, the density of levels is low, allowing quantum effects to be detected. To this end, we estimate the distance between the levels around E_n from

$$\Delta E_n = \frac{d E_n}{dn} \Delta n \qquad \text{with } \Delta n = 1.$$

We obtain

$$\Delta E_n \simeq \frac{2}{3} E_1 \, n^{-\frac{1}{3}} \simeq 10^{-11} \cdot 10^{-22} = 10^{-33} \, J,$$

which is extremely small compared to $E_n = 9.8 \ J$.

10.3   Infinite Potential Well

Determine the energy spectrum in WKB approximation for the potential

$$V(x) = V_0 \cot^2 \frac{\pi x}{a},$$

where a is a positive constant. Discuss the limits for small and large quantum numbers.

Solution

This potential is shown in Fig. 10.2.
The energy eigenvalues E_n can be determined by the relationship

$$I = \pi \hbar \left(n + \frac{1}{2}\right) \qquad \text{with } n = 0, 1, 2 \ldots, \tag{10.2}$$

where, taking into account the potential symmetry,

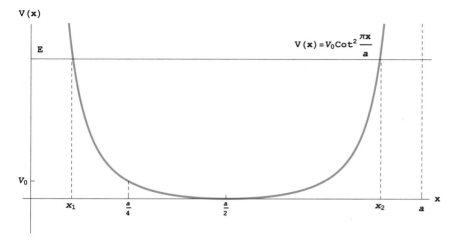

Fig. 10.2 Potential well $V(x) = V_0 \cot^2 \frac{\pi x}{a}$

$$I = 2 \int_{x_1}^{\frac{a}{2}} dx \sqrt{2m V_0} \sqrt{\frac{E}{V_0} - \cot^2 \frac{\pi x}{a}} = \frac{2a}{\pi} \sqrt{2m V_0} \int_{\frac{\pi x_1}{a}}^{\frac{\pi}{2}} dy \sqrt{\alpha - \cot^2 y},$$

and, from the turning point condition, we get

$$V(x_1) = V_0 \cot^2 \frac{\pi x_1}{a} = E_n, \quad \text{i.e.,} \quad \alpha = \frac{E_n}{V_0} = \cot^2 \frac{\pi x_1}{a}.$$

We make the following substitution:

$$z = \sqrt{\frac{\alpha - \cot^2 y}{\cot^2 y}} = \sqrt{\alpha \tan^2 y - 1}.$$

From

$$z^2 = \alpha \tan^2 y - 1 \quad \Rightarrow \quad \tan y = \sqrt{\frac{1 + z^2}{\alpha}},$$

$$z^2 = \alpha \frac{1 - \cos^2 y}{\cos^2 y} - 1 \quad \Rightarrow \quad \cos^2 y = \frac{\alpha}{z^2 + 1 + \alpha},$$

we obtain

$$dz = \frac{\alpha}{z} \frac{\tan y}{\cos^2 y} dy \quad \Rightarrow \quad dy = \frac{z}{\alpha} \frac{\cos^2 y}{\tan y} dz = \frac{z}{\alpha} \frac{\alpha}{z^2 + 1 + \alpha} \sqrt{\frac{\alpha}{1 + z^2}} dz,$$

and

$$\sqrt{\alpha - \cot^2 y} = \frac{z}{\tan y} = z\sqrt{\frac{\alpha}{1 + z^2}}.$$

The new integration limits are given by

$$y_1 = \frac{\pi x_1}{a} \qquad \Rightarrow \qquad z_1 = \sqrt{\frac{\alpha - \cot^2 \frac{\pi x_1}{a}}{\cot^2 \frac{\pi x_1}{a}}} = 0,$$

$$y_2 = \frac{\pi}{2} \qquad \Rightarrow \qquad z_2 = \sqrt{\frac{\alpha - \cot^2 \frac{\pi}{2}}{\cot^2 \frac{\pi}{2}}} = +\infty.$$

The substitution in the integral to be calculated gives us

$$\begin{aligned}
I &= \frac{2a}{\pi}\sqrt{2mV_0} \int_0^{+\infty} dz\, \frac{\alpha z^2}{(z^2 + 1 + \alpha)(z^2 + 1)} = \\
&= \frac{2a}{\pi}\sqrt{2mV_0} \int_0^{+\infty} dz\, \left[\frac{1 + \alpha}{z^2 + 1 + \alpha} - \frac{1}{z^2 + 1}\right] = \\
&= \frac{2a}{\pi}\sqrt{2mV_0}\, (\sqrt{1 + \alpha} - 1)\, \frac{\pi}{2},
\end{aligned}$$

where we used the result

$$\int_0^{+\infty} dz\, \frac{1}{z^2 + 1} = [\arctan z]_0^{+\infty} = \frac{\pi}{2}.$$

We can now impose the WKB quantization condition (10.2) and, remembering that $\alpha = \frac{E_n}{V_0}$, we obtain the energy spectrum

$$E_n = \frac{\pi^2 \hbar^2}{2ma^2}\left(n + \frac{1}{2}\right)^2 + 2\pi\hbar\sqrt{\frac{V_0}{2ma^2}}\left(n + \frac{1}{2}\right).$$

This energy spectrum for V_0 small recovers the spectrum of the square potential well and, for V_0 large, that of the harmonic oscillator (notice that V_0 is the value assumed by the potential in $\frac{a}{4}$, as shown in Fig. 10.2).

10.4   Triangular Barrier

Evaluate the WKB approximation to the probability of a particle being transmitted through the triangular barrier

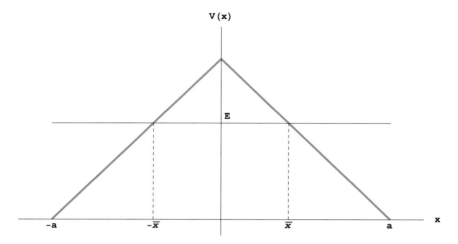

Fig. 10.3 Triangular potential barrier

$$V(x) = \begin{cases} 0, & \text{if } |x| > a; \\ V_0(1 - \frac{|x|}{a}), & \text{if } |x| < a \end{cases}.$$

for energy lower than V_0.

Solution

With reference to Fig. 10.3 the classical turning points are given by $\pm \bar{x}$, where

$$V(\bar{x}) = E \quad \Rightarrow \quad \bar{x} = a\left(1 - \frac{E}{V_0}\right).$$

Exploiting the potential symmetry, we obtain

$$\ln T = -\frac{4}{\hbar} \int_0^{\bar{x}} dx \sqrt{2m V_0} \sqrt{\left(1 - \frac{x}{a}\right) - \frac{E}{V_0}} = -\frac{4}{\hbar} \sqrt{2m V_0} \int_0^{\bar{x}} dx \sqrt{\frac{\bar{x}}{a} - \frac{x}{a}}.$$

After the substitution $y = \frac{\bar{x}}{a} - \frac{x}{a}$, we get

$$\ln T = -\frac{4a}{\hbar} \sqrt{2m V_0} \int_0^{\frac{\bar{x}}{a}} dy \sqrt{y} = -\frac{4a}{\hbar} \sqrt{2m V_0} \frac{2}{3} \left(\frac{\bar{x}}{a}\right)^{\frac{3}{2}} = -\frac{8}{3} \frac{a\sqrt{2m}}{\hbar V_0} (V_0 - E)^{\frac{3}{2}}.$$

Ultimately, the probability of transmission beyond the barrier is

$$T = e^{-\frac{8}{3} \frac{a\sqrt{2m}}{\hbar V_0} (V_0 - E)^{\frac{3}{2}}}.$$

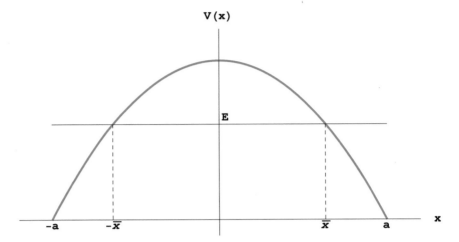

Fig. 10.4 Parabolic potential barrier

10.5   Parabolic Barrier

Use the WKB approximation method to evaluate the probability of a particle being transmitted through the parabolic barrier

$$V(x) = \begin{cases} 0, & \text{if } |x| > a; \\ V_0(1 - \frac{x^2}{a^2}), & \text{if } |x| < a \end{cases}.$$

for energy lower than V_0.

Solution

With reference to Fig. 10.4 the classical turning points are given by $\pm \bar{x}$, where

$$V(\bar{x}) = E \quad \Rightarrow \quad \bar{x} = a\sqrt{1 - \frac{E}{V_0}}.$$

Exploiting the potential symmetry, we obtain

$$\ln T = -\frac{4}{\hbar}\sqrt{2mV_0} \int_0^{\bar{x}} dx \sqrt{\left(1 - \frac{x^2}{a^2}\right) - \frac{E}{V_0}} = -\frac{4}{\hbar}\sqrt{2mV_0} \int_0^{\bar{x}} dx \sqrt{\frac{\bar{x}^2}{a^2} - \frac{x^2}{a^2}} =$$

$$= -\frac{4}{\hbar}\sqrt{2mV_0}\, \frac{\bar{x}}{a} \int_0^{\bar{x}} dx \sqrt{1 - \frac{x^2}{\bar{x}^2}}.$$

After the substitution $x = \bar{x} \sin \theta$, we get

$$\ln T = -\frac{4}{\hbar} \sqrt{2m V_0} \frac{\bar{x}^2}{a} \int_0^{\frac{\pi}{2}} d\theta \, \cos^2 \theta = -\frac{a\pi}{\hbar \sqrt{2m V_0}} \, p_E^2,$$

where $p_E = \sqrt{2m(V_0 - E)}$ is the classical momentum corresponding to a constant potential equal to V_0.

Ultimately, the probability of transmission beyond the barrier is

$$T = e^{-\frac{a\pi}{\hbar \sqrt{2m V_0}} \, p_E^2}.$$

Chapter 11
Variational Method

11.1 Ground State of an Anharmonic Oscillator

Use the Variational method to evaluate an approximation to the energy of the ground state of a quartic oscillator, i.e., a particle of mass m subject to the potential:

$$V(x) = \mu x^4 \qquad (\mu > 0).$$

Notice that this potential, like the harmonic one, diverges to infinity, has a minimum at $x = 0$ and is parity-invariant. As a consequence, the energy spectrum is discrete and the ground state has even parity and no nodes. We suggest, as trial functions that meet these requirements:

$$\psi(x; \alpha) = \text{constant} \cdot e^{-\alpha \frac{x^2}{2}},$$

the same set as the harmonic oscillator ground state (in this case, $\alpha = \frac{m\omega}{\hbar}$).

Solution

The energy expectation value in the state ψ is

$$
\begin{aligned}
E(\alpha) &= \frac{\langle \psi | \mathcal{H} | \psi \rangle}{\langle \psi | \psi \rangle} = \frac{\int_{-\infty}^{+\infty} dx\, e^{-\alpha \frac{x^2}{2}} \left[-\frac{\hbar^2}{2m} \frac{d^2}{dx^2} + \mu x^4 \right] e^{-\alpha \frac{x^2}{2}}}{\int_{-\infty}^{+\infty} dx\, e^{-\alpha x^2}} = \\
&= \sqrt{\frac{\alpha}{\pi}} \int_{-\infty}^{+\infty} dx \left[-\frac{\hbar^2}{2m}(-\alpha + \alpha^2 x^2) + \mu x^4 \right] e^{-\alpha x^2} = \\
&= \sqrt{\frac{\alpha}{\pi}} \left[\frac{\hbar^2 \alpha}{2m} I_0 - \frac{\hbar^2 \alpha^2}{2m} I_2 + \mu I_4 \right] = \frac{\hbar^2 \alpha}{4m} + \frac{3\mu}{4\alpha^2},
\end{aligned}
$$

where results (A.3) have been used:

© Springer Nature Switzerland AG 2019
L. Angelini, *Solved Problems in Quantum Mechanics*, UNITEXT for Physics,
https://doi.org/10.1007/978-3-030-18404-9_11

$$I_0 = \sqrt{\frac{\pi}{\alpha}}, \qquad I_2 = \frac{1}{2}\sqrt{\frac{\pi}{\alpha^3}}, \qquad I_4 = \frac{3}{4}\sqrt{\frac{\pi}{\alpha^5}}.$$

Now we look for the minimum of $E(\alpha)$:

$$\frac{dE(\alpha)}{d\alpha} = 0 \quad \Rightarrow \quad \frac{\hbar^2}{4m} - \frac{3\mu}{2\alpha^3} = 0 \quad \Rightarrow \quad \alpha = \sqrt[3]{\frac{6\mu m}{\hbar^2}}.$$

Notice that the second derivative is always positive, so α is effectively a minimum point. For this value of α, we obtain the following approximate value for the ground state energy:

$$E = \frac{\hbar^2 \alpha}{4m}\left(1 + \frac{3m\mu}{\hbar^2\alpha^3}\right) = \frac{\hbar^2 \alpha}{4m}\left(1 + \frac{1}{2}\right) = \frac{3\hbar^2}{8m}\sqrt[3]{\frac{6m\mu}{\hbar^2}} = \frac{3}{8}\sqrt[3]{\frac{6\mu\hbar^4}{m^2}}.$$

11.2   Ground State of a Potential Well

Use the variational method to evaluate the energy of the ground state for a particle bound on the segment $[-a, a]$. Use, as a trial function,

$$\psi(x; \alpha) = |x|^\alpha - a^\alpha,$$

where α is the variational parameter. Compare this approximation with the exact value and determine the relative error.

Solution

The energy expectation value in the state ψ is

$$E(\alpha) = \frac{\langle\psi|\mathcal{H}|\psi\rangle}{\langle\psi|\psi\rangle} = \frac{2\int_0^{+a} dx\,(x^\alpha - a^\alpha)\left[-\frac{\hbar^2}{2m}\frac{d^2}{dx^2}\right](x^\alpha - a^\alpha)}{2\int_0^{+a} dx\,(x^\alpha - a^\alpha)^2} =$$

$$= -\frac{\hbar^2}{4ma^2}\frac{1 + 3\alpha + 2\alpha^2}{1 - 2\alpha}.$$

By requiring that the derivative of $E(\alpha)$ vanishes, there are two solutions

$$\alpha_\pm = \frac{1}{2}\left(1 \pm \sqrt{6}\right).$$

The solution relative to the $+$ sign gives us the minimum point sought (the other solution corresponds to a maximum point). For this value of α, we obtain the following approximate value for the ground state energy:

$$E_1^{var} = \frac{1.23737\,\hbar^2}{ma^2},$$

to be compared with the exact value

$$E_1 = \frac{\pi^2\hbar^2}{8ma^2}.$$

The relative error is equal to

$$\frac{E_1^{var} - E_1}{E_1} = 0.002976,$$

less, therefore, than 0.3%.

11.3 First Energy Levels of a Linear Potential

A particle of mass m moves into a linear central potential

$$V(r) = Kr.$$

For this potential, the radial Schrödinger equation presents a natural scale of lengths $r_0 = \sqrt[3]{\frac{\hbar^2}{2mk}}$; in terms of the scaled position $\rho = \frac{r}{r_0}$, it can be rewritten as

$$\left[-\frac{d^2}{d\rho^2} + \frac{\ell(\ell+1)}{\rho^2} - \rho \right] \psi(\rho) = \varepsilon\psi(\rho),$$

where

$$\varepsilon = \frac{E}{E_0} \quad \text{with} \quad E_0 = \sqrt[3]{\frac{\hbar^2 k^2}{2m}}.$$

By numerically solving this equation, which, in the case $\ell = 0$, is the Airy equation, you find that the lowest eigenvalues for $\ell = 0, 1, 2$ are, respectively, $\varepsilon = 2.3381, 3.3612, 4.2482$. Exploiting a trial function that has the correct trend for $\rho \to 0$, use the variational method to evaluate these eigenvalues.

Solution

Due to the centrifugal potential, the wave function behaves in the origin as $\rho^{\ell+1}$. Because the potential diverges asymptotically, we can use the trial function

$$\psi(\rho) = \rho^{\ell+1} e^{-\alpha\rho}.$$

The energy expectation value in this state is

$$E(\alpha) = \frac{\langle \psi | \mathcal{H} | \psi \rangle}{\langle \psi | \psi \rangle} = \frac{\int_0^\infty d\rho\, \rho^{\ell+1}\, e^{-\alpha\rho}\left[-\frac{d^2}{d\rho^2} + \frac{\ell(\ell+1)}{\rho^2} - \rho\right]\rho^{\ell+1}\, e^{-\alpha\rho}}{\int_0^\infty d\rho\, \rho^{2(\ell+1)}\, e^{-2\alpha\rho}} =$$

$$= \frac{(2\ell+3)!\,(2\alpha)^{-2(\ell+2)}(2\alpha^3 + 2\ell + 3)}{(2\ell+3)!\,(2\alpha)^{-(2\ell+3)}} = \frac{2\alpha^3 + 2\ell + 3}{2\alpha},$$

where, to calculate the integrals, (A.6) has been used. As a function of α, $E(\alpha)$ has a minimum that can be calculated by imposing the zeroing of its derivative:

$$\frac{4\alpha^3 - 2\ell - 3}{2\alpha^2} = 0 \qquad \rightarrow \qquad \alpha(\ell) = \sqrt[3]{\frac{3 + 2\ell}{4}}.$$

Substituting, in $E(\alpha)$, the values $\ell = 0, 1, 2$, the following results are obtained:

$$E(\alpha(0)) = 2.4764, \qquad E(\alpha(1)) = 3.4812, \qquad E(\alpha(2)) = 4.3566,$$

which reproduce the numerical results with a variable precision between 3 and 6%. For the test wave function, an asymptotic Gaussian pattern could also be used. The integrals are equally calculable, using this time (A.2), and results are obtained with a slightly higher precision.

11.4 Ground State of Helium

The Helium atom has atomic number $Z = 2$ and mass number $A = 4$; thus, there are two electrons moving around a nucleus of mass equal to 4 times the mass of Hydrogen and about 8×10^3 the mass of an electron. Assuming this mass to be infinite, we refer to the coordinates of the two electrons with respect to the nucleus with $\mathbf{r}_1$ and $\mathbf{r}_2$ and to their relative position with $\mathbf{r}_{12} = \mathbf{r}_1 - \mathbf{r}_2$. The Helium atom Hamiltonian is, therefore,

$$\mathcal{H} = -\frac{\hbar^2}{2m}(\nabla_1^2 + \nabla_2^2) - \frac{Ze^2}{r_1} - \frac{Ze^2}{r_2} + \frac{e^2}{r_{12}}.$$

If there were no repulsive interaction between the two electrons, in the Schrödinger equation, there would be a separation between the variables of the two electrons and, E being the energy of the ground state, we would have

$$\psi_E^0(\mathbf{r}_1, \mathbf{r}_2) = \psi_{E_1}(\mathbf{r}_1)\,\psi_{E_1}(\mathbf{r}_2) = \frac{Z^3}{\pi a_0^3}\, e^{-\frac{Z(r_1 + r_2)}{a_0}}. \tag{11.1}$$

Here, we denoted by E_1 and $\psi_{E_1}(r) = (2/\sqrt{4\pi})(Z/a_0)^{\frac{3}{2}} \cdot e^{-\frac{Zr}{a_0}}$, respectively, the ground state energy of a Hydrogenlike atom with $Z = 2$ and the corresponding eigenfunction. In this very rough approximation, neglecting the positive term of attraction between the electrons, the ground state energy would be

$$E^0 = 2\,E_1 = 2\left[-\frac{m(Ze^2)^2}{2\hbar^2}\right] = -8 \cdot 13.6\,eV = -108.8\,eV.$$

Experimentally, it is found that the Helium ground state has energy $E = -78.98\,eV$, a much higher value.

Use the Variational method to improve this approximation.

Hint: the trial functions will again be given by (11.1), however, considering Z as a parameter to be determined by minimizing the functional

$$E(Z) = \left(\frac{Z^3}{\pi a_0^3}\right)^2 \int d\mathbf{r}_1 d\mathbf{r}_2\, e^{-\frac{Z(r_1+r_2)}{a_0}} \mathcal{H} e^{-\frac{Z(r_1+r_2)}{a_0}}.$$

Solution

The Hamiltonian can be rewritten in the form

$$\mathcal{H} = -\frac{\hbar^2}{2m}(\nabla_1^2 + \nabla_2^2) - \frac{Ze^2}{r_1} - \frac{(2-Z)e^2}{r_1} - \frac{Ze^2}{r_2} - \frac{(2-Z)e^2}{r_2} + \frac{e^2}{r_{12}}.$$

Taking into account that ψ_{E_1} is the solution of the Schrödinger equation in correspondence with the eigenvalue E_1, i.e.,

$$\left(-\frac{\hbar^2}{2m}\nabla_1^2 - \frac{Ze^2}{r_1}\right)e^{-\frac{Zr_1}{a_0}} = -\frac{Z^2e^2}{2a_0}\,e^{-\frac{Zr_1}{a_0}},$$

it is possible to eliminate the differential operators and so we find

$$E(Z) = \left(\frac{Z^3}{\pi a_0^3}\right)^2 \int d\mathbf{r}_1 d\mathbf{r}_2\, e^{-\frac{2Z(r_1+r_2)}{a_0}} e^2 \left[-\frac{Z^2}{a_0} - \frac{(2-Z)}{r_1} - \frac{(2-Z)}{r_2} + \frac{1}{r_{12}}\right] =$$

$$= \left(\frac{Z^3 e}{\pi a_0^3}\right)^2 \left[-\frac{Z^2}{a_0}K_1 - 2(2-Z)K_2 + K_3\right].$$

Here, K_1, K_2 and K_3 are the following integrals:

- K_1

$$K_1 = \int d\mathbf{r}_1 d\mathbf{r}_2\, e^{-\frac{2Z(r_1+r_2)}{a_0}} = \left(4\pi \int_0^\infty dr\, r^2 e^{-\beta r}\right)^2 = \left(4\pi \frac{2}{\beta^3}\right)^2,$$

where $\beta = \frac{2Z}{a_0}$ and (A.8) have been used.

- K_2

$$K_2 = \int d\mathbf{r_1} d\mathbf{r_2} \, e^{-\beta(r_1+r_2)} \frac{1}{r_1} = (4\pi)^2 \int_0^\infty dr_2 \, r_2^2 \, e^{-\beta r_2} \int_0^\infty dr_1 \, r_1 e^{-\beta r_1} =$$

$$= (4\pi)^2 \frac{2}{\beta^3} \frac{1}{\beta^2} = \frac{2(4\pi)^2}{\beta^5}.$$

- $\mathbf{K_3}$ (this calculation is also present in Problem 6.27)

$$K_3 = \int d\mathbf{r_1} d\mathbf{r_2} \, e^{-\beta(r_1+r_2)} \frac{1}{|\mathbf{r_1} - \mathbf{r_2}|} = \int d\mathbf{r_1} e^{-\beta r_1} f(r_1),$$

where

$$f(r_1) = \int d\mathbf{r_2} \, e^{-\beta r_2} \frac{1}{|\mathbf{r_1} - \mathbf{r_2}|}$$

is a function of $\mathbf{r_1}$ only, because, by integrating over the whole $\mathbf{r_2}$ solid angle, the dependence on the angles disappears. To calculate $f(r_1)$, we are free, therefore, to position $\mathbf{r_1}$ in the direction of the z axis.
We get

$$f(r_1) = 2\pi \int dr_2 \, r_2^2 \, e^{-\beta r_2} \int_{-1}^{+1} d\cos\theta \, \frac{1}{\sqrt{r_1^2 + r_2^2 - 2r_1 r_2 \cos\theta}} =$$

$$= -2\pi \int dr_2 \, r_2^2 \, e^{-\beta r_2} \frac{1}{r_1 r_2} \left[\sqrt{r_1^2 + r_2^2 - 2r_1 r_2 \cos\theta} \right]_{-1}^{+1} =$$

$$= \frac{2\pi}{r_1} \int dr_2 \, r_2 \, e^{-\beta r_2} (r_1 + r_2 - |r_1 - r_2|) =$$

$$= \frac{2\pi}{r_1} \left[\int_0^{r_1} dr_2 \, r_2 \, e^{-\beta r_2} (2r_2) + \int_{r_1}^\infty dr_2 \, r_2 \, e^{-\beta r_2} (2r_1) \right] =$$

$$= \frac{4\pi}{r_1 \beta^3} \left[2 - e^{-\beta r_1} (2 + \beta r_1) \right],$$

where we used the expressions, derived from (A.7),

$$I_2(0, r_1) = \frac{1}{\beta^3} [2 - (2 + 2r_1\beta + r_1^2\beta^2)e^{-\beta r_1}]$$

and

$$I_1(r_1, \infty) = \frac{1}{\beta^2} (1 + \beta r_1)e^{-\beta r_1}.$$

Returning to the calculation of K_3, we obtain, using the usual integrals (A.8),

$$K_3 = 4\pi \int_0^\infty dr_1 \, r_1^2 \, e^{-\beta r_1} \frac{4\pi}{r_1 \beta^3} \left[2 - e^{-\beta r_1} (2 + \beta r_1) \right] = \frac{(4\pi)^2}{\beta^5} \frac{5}{4}.$$

By inserting these results into the expression for $E(Z)$, we obtain

$$E(Z) = -\frac{e^2}{a_0}\left(\frac{27}{8}Z - Z^2\right),$$

which takes its minimum value for

$$Z = Z_{eff} = \frac{27}{16}$$

given by

$$E(Z_{eff}) = -\frac{e^2}{a_0}\left(\frac{27}{8}Z_{eff} - Z_{eff}^2\right) = -77.5 \, eV.$$

This value is very close to the already mentioned experimental value $-78.98\,eV$ and provides a better approximation compared to the calculation in Perturbation Theory (see Problem 6.27). Both approximations calculate the energy of the ground state as the expectation value of the Hamiltonian in the unperturbed state, but the variational method modifies the value of Z in such a way as to minimize the difference from the exact value.

Notice, moreover, that

$$Z_{eff} < 2.$$

Taking into account that wave functions are factored, we can consider the approximation as resulting from a model in which each of the electrons moves in the mean field of a nucleus with an effective charge $Z_{eff} \cdot e$, lower than the real one due to the screen effect produced by the other electron.

Appendix
Useful Formulas

A.1   Frequently Used Integrals

A.1.1   Gaussian Integrals

Having defined

$$I_0(\alpha) = \int_{-\infty}^{+\infty} dx \, e^{-\alpha x^2} = \sqrt{\frac{\pi}{\alpha}}, \tag{A.1}$$

we have

$$I_{2n+1}(\alpha) = \int_{-\infty}^{+\infty} dx \, x^{2n+1} e^{-\alpha x^2} = 0,$$

$$I_{2n}(\alpha) = \int_{-\infty}^{+\infty} dx \, x^{2n} e^{-\alpha x^2} = (-1)^n \frac{\partial^n}{\partial \alpha^n} I_0(\alpha) = (-1)^n \frac{\partial^n}{\partial \alpha^n} \sqrt{\frac{\pi}{\alpha}}, \tag{A.2}$$

for $n = 1, 2, \ldots$.

The results for the first values of n are

$$I_0 = \sqrt{\frac{\pi}{\alpha}}, \qquad I_2 = \frac{1}{2}\sqrt{\frac{\pi}{\alpha^3}}, \qquad I_4 = \frac{3}{4}\sqrt{\frac{\pi}{\alpha^5}}. \tag{A.3}$$

Another Gaussian integral of frequent use is

$$I(\alpha, \beta) = \int_{-\infty}^{+\infty} dx \, e^{-\alpha x^2 + \beta x} = \sqrt{\frac{\pi}{\alpha}} \, e^{\frac{\beta^2}{4\alpha}}, \tag{A.4}$$

which also allows you to calculate

© Springer Nature Switzerland AG 2019
L. Angelini, *Solved Problems in Quantum Mechanics*, UNITEXT for Physics,
https://doi.org/10.1007/978-3-030-18404-9

$$I(\alpha, \beta) = \int_0^{+\infty} dx \, x \, \sin(\beta x) \, e^{-\alpha^2 x^2} =$$

$$= \frac{1}{2} \int_{-\infty}^{+\infty} dx \, x \, \frac{e^{\iota \beta x} - e^{-\iota \beta x}}{2\iota} \, e^{-\alpha^2 x^2} =$$

$$= -\frac{1}{4} \frac{\partial}{\partial \beta} \int_{-\infty}^{+\infty} dx \, \frac{e^{\iota \beta x} + e^{-\iota \beta x}}{\iota} \, e^{-\alpha^2 x^2} =$$

$$= -\frac{1}{4} \frac{\partial}{\partial \beta} e^{-\frac{\beta^2}{4\alpha^2}} \int_{-\infty}^{+\infty} dx \left[e^{-(\alpha x + \iota \frac{\beta}{2\alpha})^2} + e^{-(\alpha x - \iota \frac{\beta}{2\alpha})^2} \right] =$$

$$= -\frac{1}{4} \frac{\partial}{\partial \beta} e^{-\frac{\beta^2}{4\alpha^2}} 2 \frac{\sqrt{\pi}}{\alpha} =$$

$$= \frac{\sqrt{\pi} \beta}{4\alpha^3} e^{-\frac{\beta^2}{4\alpha^2}}. \tag{A.5}$$

A.1.2   Integrals of Exponential Functions and Powers

$$I_n(0, \infty) = \int_0^{+\infty} dx \, x^n e^{-x} = \left[(-1)^n \frac{d^n}{d\alpha^n} \int_0^{+\infty} dx \, e^{-\alpha x} \right]_{\alpha=1} =$$

$$= \left[(-1)^n \frac{d^n}{d\alpha^n} \frac{1}{\alpha} \right]_{\alpha=1} = n! \tag{A.6}$$

Similarly, we find the most general result is

$$I_n(a, b) = \int_a^b dx \, x^n \, e^{-\beta x} = (-1)^n \frac{d^n}{d\beta^n} I_0(a, b). \tag{A.7}$$

For $a = 0$ and $b = \infty$, we get

$$I_0(0, \infty) = \frac{1}{\beta}; \quad I_1(0, \infty) = -\frac{d}{d\beta} \frac{1}{\beta} = \frac{1}{\beta^2}; \quad I_2(0, \infty) = \frac{d^2}{d\beta^2} \frac{1}{\beta} = \frac{2}{\beta^3}. \tag{A.8}$$

A.2   Continuity Equation

The continuity equation in Quantum Mechanics is

$$\frac{\partial P(\mathbf{r}, t)}{\partial t} = -\nabla \cdot \mathbf{j}(\mathbf{r}, t), \tag{A.9}$$

where the probability density P is given by

$$P(\mathbf{r}, t) = |\psi(\mathbf{r}, t)|^2 = \psi^*(\mathbf{r}, t)\,\psi(\mathbf{r}, t) \tag{A.10}$$

and the probability current density is defined as

$$\mathbf{j}(\mathbf{r}, t) = \frac{\hbar}{2im}\left[\psi^*(\mathbf{r}, t)\nabla\psi(\mathbf{r}, t) - \psi(\mathbf{r}, t)\nabla\psi^*(\mathbf{r}, t)\right] = \frac{\hbar}{im}\,\Im(\psi^*(\mathbf{r}, t)\nabla\psi(\mathbf{r}, t)). \tag{A.11}$$

A.3 Harmonic Oscillator

A.3.1 Operator Treatment

The eigenvalues of the harmonic oscillator Hamiltonian are

$$E_n = (n + \frac{1}{2})\hbar\omega, \qquad \hat{H}|n\rangle = E_n|n\rangle. \tag{A.12}$$

In terms of raising and lowering (creation and destruction) operators,

$$a^\dagger = \sqrt{\frac{m\omega}{2\hbar}}\,x - i\sqrt{\frac{1}{2m\omega\hbar}}\,p, \quad a = \sqrt{\frac{m\omega}{2\hbar}}\,x + i\sqrt{\frac{1}{2m\omega\hbar}}\,p, \tag{A.13}$$

the position and momentum operators are given by

$$x = \sqrt{\frac{\hbar}{2m\omega}}\,(a + a^\dagger), \qquad p = \frac{1}{i}\sqrt{\frac{\hbar m\omega}{2}}\,(a - a^\dagger). \tag{A.14}$$

a and a^+ act on the energy eigenkets as follows

$$a|n\rangle = \sqrt{n}\,|n-1\rangle \qquad a^+|n\rangle = \sqrt{n+1}\,|n+1\rangle. \tag{A.15}$$

A.3.2 Position Basis Treatment

The eigenfunctions in the position basis are given by

$$\phi_n(x) = \left(\frac{m\omega}{\pi\hbar}\right)^{\frac{1}{4}}\frac{1}{\sqrt{2^n n!}}\,e^{-\frac{\xi^2}{2}}\,H_n(\xi), \qquad \text{where} \qquad \xi = \sqrt{\frac{m\omega}{\hbar}}x, \tag{A.16}$$

where H_n is the Hermite polynomial defined by

$$H_n(\xi) = (-1)^n e^{\xi^2} \frac{d^n e^{-\xi^2}}{d\xi^n}. \tag{A.17}$$

Hermite polynomials are orthogonal polynomials,

$$\int_{-\infty}^{+\infty} d\xi \, H_n(\xi) H_m(\xi) = \sqrt{\pi} \, 2^n \, n! \, \delta_{n,m} \tag{A.18}$$

and they satisfy the following recurrence relation:

$$2\xi \, H_n(\xi) = H_{n+1}(\xi) + 2n \, H_{n-1}(\xi). \tag{A.19}$$

First Hermite polynomials

$$H_0(x) = 1, \quad H_1(x) = 2x, \quad H_2(x) = 4x^2 - 2, \quad H_3(x) = 8x^3 - 12x,$$
$$H_4(x) = 16x^4 - 48x^2 + 12, \quad H_5(x) = 32x^5 - 160x^3 + 120x. \tag{A.20}$$

A.4   Spherical Coordinates

The transition from Cartesian coordinates to spherical coordinates occurs through the transformation:

$$x = r \sin\theta \cos\phi, \tag{A.21}$$
$$y = r \sin\theta \sin\phi, \tag{A.22}$$
$$z = r \cos\theta. \tag{A.23}$$

A.5   Angular Momentum

A.5.1   Operator Treatment

The operators J^2, J_x, J_y, J_z satisfy the following commutation relations:

$$[J^2, J_x] = [J^2, J_y] = [J^2, J_z] = 0,$$

$$[J_x, J_y] = i\hbar J_z, \qquad [J_y, J_z] = i\hbar J_x, \qquad [J_z, J_x] = i\hbar J_y.$$

The J^2 and J_z common basis is denoted by $|j, m\rangle$:

$$J^2|j, m\rangle = j(j + 1)\hbar^2|j, m\rangle, \qquad J_z|j, m\rangle = m\hbar|j, m\rangle.$$

The operators

$$J_\pm = J_x \pm i J_y \qquad (A.24)$$

satisfy the following commutation relations with the operators J^2 and J_z:

$$[J^2, J_\pm] = 0, \qquad [J_z, J_\pm] = \pm J_\pm. \qquad (A.25)$$

$J_\pm$ act on an eigenket common to J^2 and J_z, raising or lowering the azimuthal quantum number:

$$J_\pm|j, m\rangle = \hbar\sqrt{j(j + 1) - m(m \pm 1)}\ |j, m \pm 1\rangle. \qquad (A.26)$$

A.5.2 Spherical Harmonics

Definition

$$Y_{\ell,m}(\theta, \phi) = (-1)^m \sqrt{\frac{2\ell + 1}{4\pi} \frac{(\ell - |m|)!}{(\ell + |m|)!}}\ P_\ell^m(\cos\theta)\, e^{im\phi}, \qquad (A.27)$$

where P_ℓ^m are the Legendre associate functions defined for $|z| \le 1$,

$$P_\ell^m(z) = (1 - z^2)^{\frac{|m|}{2}} \frac{d^{|m|}}{dz^{|m|}} P_\ell(z), \qquad (A.28)$$

which, for $m = 0$, give us the Legendre polynomials $P_\ell(z)$,

$$P_\ell(z) = \frac{1}{2^\ell \ell!} \frac{d^\ell}{dz^\ell} (1 - z^2)^\ell. \qquad (A.29)$$

They are orthogonal polynomials:

$$\int_{-1}^{+1} dz\, P_\ell(z) P_{\ell'}(z) = \frac{2}{2\ell + 1} \delta_{\ell,\ell'}. \qquad (A.30)$$

Particular values of Legendre polynomials and associate functions:

$$P_\ell(\pm 1) = (\pm 1)^\ell, \qquad P_\ell^m(\pm 1) = 0 \text{ for } m \ne 0. \qquad (A.31)$$

First Legendre polynomials:

$$P_0(z) = 1, \quad P_1(z) = z, \quad P_2(z) = \frac{1}{2}(3z^2 - 1), \tag{A.32}$$

$$P_3(z) = \frac{1}{2}(5z^3 - 3z), \quad P_4(z) = \frac{1}{8}(35z^4 - 30z^2 + 3). \tag{A.33}$$

Orthonormalization relationship

$$\int d\Omega \, Y^*_{\ell',m'}(\theta, \phi) \, Y_{\ell,m}(\theta, \phi) = \delta_{\ell,\ell'} \, \delta_{m,m'}. \tag{A.34}$$

Recurrence relationship

$$\cos\theta \, Y_{\ell,m}(\theta, \phi) = a_{\ell,m} \, Y_{\ell+1,m}(\theta, \phi) + a_{\ell-1,m} \, Y_{\ell-1,m}(\theta, \phi), \tag{A.35}$$

where

$$a_{\ell,m} = \sqrt{\frac{(\ell+1+m)(\ell+1-m)}{(2\ell+1)(2\ell+3)}}. \tag{A.36}$$

Sum theorem

If (Θ, Φ) and (θ', ϕ') are two space directions and θ is the angle between them, a Legendre polynomial can be expressed in terms of spherical harmonics:

$$P_\ell(\cos\theta) = \frac{4\pi}{2\ell+1} \sum_{m=-\ell}^{+\ell} Y_{\ell,m}(\Theta, \Phi)^* \, Y_{\ell,m}(\theta', \phi'). \tag{A.37}$$

First Spherical Harmonics

$$Y_{0,0}(\theta, \phi) = \frac{1}{\sqrt{4\pi}}, \tag{A.38}$$

$$Y_{1,0}(\theta, \phi) = \sqrt{\frac{3}{4\pi}} \cos\theta, \quad Y_{1,\pm1}(\theta, \phi) = \mp\sqrt{\frac{3}{8\pi}} \sin\theta e^{\pm i\phi}, \tag{A.39}$$

$$Y_{2,0}(\theta, \phi) = \sqrt{\frac{5}{16\pi}}(3\cos^2\theta - 1), \quad Y_{2,\pm1}(\theta, \phi) = \mp\sqrt{\frac{15}{8\pi}} \sin\theta \cos\theta e^{\pm i\phi},$$

$$Y_{2,\pm2}(\theta, \phi) = \sqrt{\frac{15}{32\pi}} \sin^2\theta e^{\pm 2i\phi}. \tag{A.40}$$

$$Y_{3,0}(\theta, \phi) = \sqrt{\frac{7}{16\pi}}(5\cos^3\theta - 3\cos\theta) \; , \quad Y_{3,\pm1}(\theta, \phi) = \mp\sqrt{\frac{21}{64\pi}}\sin\theta(5\cos^2\theta - 1)e^{\pm\iota\phi},$$

$$Y_{3,\pm2}(\theta, \phi) = \sqrt{\frac{105}{32\pi}}\sin^2\theta\cos\theta e^{\pm2\iota\phi} \; , \quad Y_{3,\pm3}(\theta, \phi) = \mp\sqrt{\frac{35}{64\pi}}\sin^3\theta e^{\pm3\iota\phi}. \tag{A.41}$$

A.6 Schrödinger Equation in Spherical Coordinates

A.6.1 Radial Equation

If the potential energy $V(r)$ is central, the Schrödinger equation is separable into spherical coordinates. The eigenfunction common to the operators $\mathcal{H}$, L^2 and L_z, with eigenvalues E, $\ell(\ell + 1)\hbar^2$ and $m\hbar$, respectively, can be written in the form

$$\psi_{E,\ell,m}(r, \theta, \phi) = R_{E,\ell}(r) Y_{\ell,m}(\theta, \phi) = \frac{U_{E,\ell}(r)}{r} Y_{\ell,m}(\theta, \phi), \tag{A.42}$$

where $U_{E,\ell}(r)$ is the solution to the radial equation:

$$-\frac{\hbar^2}{2m}\frac{d^2 U_{E,\ell}}{dr^2} + \frac{\hbar^2\ell(\ell + 1)}{2mr^2}U_{E,\ell} + V(r)U_{E,\ell} = E\,U_{E,\ell}, \tag{A.43}$$

with m reduced mass of the system.
$U_{E,\ell}(r)$ must satisfy the condition

$$\lim_{r \to 0} U_{E,\ell}(r) = 0. \tag{A.44}$$

A.7 Spherical Bessel Functions

The Spherical Bessel functions are solutions to the Spherical Bessel equation

$$z^2\frac{d^2}{dz^2}\phi(z) + 2z\frac{d}{dz}\phi(z) + \left[z^2 - \ell(\ell + 1)\right]\phi(z) = 0. \tag{A.45}$$

A.7.1 Spherical Bessel Functions of the First and Second Kinds

Two linearly independent integrals of (A.45) are given by the spherical Bessel functions of the first and second kinds j_ℓ and $y_\ell = (-1)^{\ell+1}j_{-\ell-1}$. For the first integer values of ℓ, they are

$$j_0(z) = \frac{\sin z}{z}, \tag{A.46}$$

$$j_1(z) = \frac{\sin z}{z^2} - \frac{\cos z}{z}, \tag{A.47}$$

$$j_2(z) = \left(\frac{3}{z^2} - 1\right)\frac{\sin z}{z} - \frac{3\cos z}{z^2}, \tag{A.48}$$

$$j_3(z) = \left(\frac{15}{z^3} - \frac{6}{z}\right)\frac{\sin z}{z} - \left(\frac{15}{z^2} - 1\right)\frac{\cos z}{z}, \tag{A.49}$$

and

$$y_0(z) = -\frac{\cos z}{z}, \tag{A.50}$$

$$y_1(z) = -\frac{\cos z}{z^2} - \frac{\sin z}{z}, \tag{A.51}$$

$$y_2(z) = \left(-\frac{3}{z^2} + 1\right)\frac{\cos z}{z} - \frac{3\sin z}{z^2}, \tag{A.52}$$

$$y_3(z) = \left(-\frac{15}{z^3} + \frac{6}{z}\right)\frac{\cos z}{z} - \left(\frac{15}{z^2} - 1\right)\frac{\sin z}{z}. \tag{A.53}$$

Their asymptotic behavior is given by

$$j_\ell(z) \underset{z\to\infty}{\sim} \frac{1}{z}\cos\left(z - \frac{\ell+1}{2}\pi\right) \tag{A.54}$$

and

$$y_\ell(z) \underset{z\to\infty}{\sim} \frac{1}{z}\sin\left(z - \frac{\ell+1}{2}\pi\right), \tag{A.55}$$

while the behavior in the origin is given by

$$j_\ell(z) \underset{z\to 0}{\sim} \frac{z^\ell}{(2\ell+1)!!} \tag{A.56}$$

and

$$y_\ell(z) \underset{z\to\infty}{\sim} -\frac{(2\ell-1)!!}{z^{\ell+1}}. \tag{A.57}$$

A.7.2 Spherical Hankel Functions

Other linearly independent solutions to the Spherical Bessel equation are the spherical Hankel functions of the first and second kinds defined by

$$h_\ell^{(1)}(z) = j_\ell(z) + \imath y_\ell(z), \tag{A.58}$$

$$h_\ell^{(2)}(z) = j_\ell(z) - \imath y_\ell(z). \tag{A.59}$$

Their asymptotic behavior is given by

$$h_\ell^{(1)}(z) \underset{z\to\infty}{\sim} \frac{1}{z} e^{\imath\left(z - \frac{\ell+1}{2}\pi\right)}, \tag{A.60}$$

$$h_\ell^{(2)}(z) \underset{z\to\infty}{\sim} \frac{1}{z} e^{-\imath\left(z - \frac{\ell+1}{2}\pi\right)}. \tag{A.61}$$

When the argument is an imaginary number, the Hankel functions have an exponential asymptotic behavior:

$$h_\ell^{(1)}(\imath z) \underset{z\to\infty}{\sim} \frac{1}{\imath z} e^{\left(-z - \imath\frac{\ell+1}{2}\pi\right)}, \tag{A.62}$$

and

$$h_\ell^{(2)}(\imath z) \underset{z\to\infty}{\sim} \frac{1}{\imath z} e^{\left(z + \imath\frac{\ell+1}{2}\pi\right)}. \tag{A.63}$$

A.8 Hydrogen Atom First Energy Eigenfunctions

Having introduced $a_0 = \frac{\hbar^2}{\mu e^2}$, the Bohr radius, the first two energy eigenfunctions are

$$\psi_{1,0,0} = \frac{1}{\sqrt{\pi}} a_0^{-\frac{3}{2}} e^{-\frac{r}{a_0}}, \tag{A.64}$$

$$\psi_{2,0,0} = \frac{1}{4\sqrt{2\pi}} a_0^{-\frac{3}{2}} \left(2 - \frac{r}{a_0}\right) e^{-\frac{r}{2a_0}}, \tag{A.65}$$

$$\psi_{2,1,0} = \frac{1}{4\sqrt{2\pi}} a_0^{-\frac{3}{2}} \frac{r}{a_0} e^{-\frac{r}{2a_0}} \cos\theta, \tag{A.66}$$

$$\psi_{2,1,\pm1} = \frac{1}{8\sqrt{2\pi}} a_0^{-\frac{3}{2}} \frac{r}{a_0} e^{-\frac{r}{2a_0}} \sin\theta \, e^{\pm\imath\varphi}. \tag{A.67}$$

A.9   Spin

A.9.1   Pauli Matrices

$$\sigma_1 = \begin{pmatrix} 0 & 1 \\ 1 & 0 \end{pmatrix} \quad , \sigma_2 = \begin{pmatrix} 0 & -i \\ i & 0 \end{pmatrix} \quad , \sigma_3 = \begin{pmatrix} 1 & 0 \\ 0 & -1 \end{pmatrix}. \tag{A.68}$$

$$\sigma_i \sigma_j = \delta_{ij} + \epsilon_{ijk}\sigma_k, \tag{A.69}$$

$$\{\sigma_i, \sigma_j\} = \sigma_i \sigma_j + \sigma_j \sigma_i = 2\delta_{ij}, \tag{A.70}$$

$$[\sigma_i \sigma_j] = \sigma_i \sigma_j - \sigma_j \sigma_i = 2i\epsilon_{ijk}\sigma_k. \tag{A.71}$$

A.9.2   Useful Relationships

$$(\mathbf{A} \cdot \boldsymbol{\sigma})\,(\mathbf{B} \cdot \boldsymbol{\sigma}) = (\mathbf{A} \cdot \mathbf{B})\,\mathbb{I} + i\,(\mathbf{A} \times \mathbf{B}) \cdot \boldsymbol{\sigma}, \tag{A.72}$$

where $\mathbb{I}$ is the identity matrix. In particular, if $\mathbf{A} = \mathbf{B}$,

$$(\mathbf{A} \cdot \boldsymbol{\sigma})^2 = A^2 \mathbb{I}, \tag{A.73}$$

$$e^{i\boldsymbol{\theta} \cdot \boldsymbol{\sigma}} = \mathbb{I} \cos\theta + i(\mathbf{n} \cdot \boldsymbol{\sigma}) \sin\theta, \quad \text{where } \mathbf{n} = \frac{\boldsymbol{\theta}}{\theta}. \tag{A.74}$$

A.10   Time-Independent Perturbation Theory

Let us consider the Hamiltonian

$$\mathcal{H} = \mathcal{H}_0 + \mathcal{H}_1,$$

where the solution to the $\mathcal{H}_0$ eigenvalue problem is supposed to be known:

$$\mathcal{H}_0 |n^0\rangle = E_n^0 |n^0\rangle.$$

If E_n^0 is not a degenerate eigenvalue and the matrix elements $\langle m^0 | \mathcal{H}_1 | n^0 \rangle$ are small compared to E_n^0, given the following expansions for the $\mathcal{H}$ eigenvalues E_n and eigenkets $|n\rangle$:

$$E_n = E_n^0 + E_n^1 + E_n^2 + \cdots ,$$

$$|n\rangle = |n^0\rangle + |n^1\rangle + |n^2\rangle + \cdots ,$$

we get

$$E_n^1 = \langle n^0 | \mathcal{H}_1 | n^0 \rangle, \tag{A.75}$$

$$E_n^2 = \sum_{m \neq n} \frac{|\langle m^0 | \mathcal{H}_1 | n^0 \rangle|^2}{E_n^0 - E_m^0}, \tag{A.76}$$

$$|n^1\rangle = \sum_{m \neq n} \frac{\langle m^0 | \mathcal{H}_1 | n^0 \rangle}{E_n^0 - E_m^0} |m^0\rangle. \tag{A.77}$$

If E_n^0 is degenerate, the first-order corrections to the eigenvalues are given from the eigenvalues of the matrix representative of $\mathcal{H}_1$ in the E_n^0 eigenspace, obtained from

$$\det \left[(\mathcal{H}_1)_{m,j} - E_n^1 \delta_{m,j} \right] = 0. \tag{A.78}$$

A.11   Sudden Perturbation

A sudden perturbation is an abrupt change of the Hamiltonian

$$\mathcal{H}_0 \quad \to \quad \mathcal{H} = \mathcal{H}_0 + \mathcal{H}_1,$$

where $\mathcal{H}_0$ and $\mathcal{H}_1$ do not depend on time. A sudden perturbation does not modify the state vector. Assuming that the system is initially in a state $|n^0\rangle$, an eigenket of $\mathcal{H}_0$, the probability of measuring an energy E_k, eigenvalue of the new Hamiltonian, that is, the probability of the transition $|n^0\rangle \to |k\rangle$, is given by

$$P_{n \to k} = |\langle k | n^0 \rangle|^2. \tag{A.79}$$

If it make sense to apply Perturbation Theory for non-degenerate eigenvalues, the probability of a transition to states $k \neq n$ is

$$P_{n \to k} = \left| \frac{\langle k^0 | \mathcal{H}_1 | n^0 \rangle}{E_k^0 - E_n^0} \right|^2. \tag{A.80}$$

A.12   Time-Dependent Perturbation Theory

Let us consider a Hamiltonian

$$\mathcal{H} = \mathcal{H}_0 + \mathcal{H}_1(t)$$

for which the solution to the $\mathcal{H}_0$ eigenvalue problem is known:

$$\mathcal{H}_0|n^0\rangle = E_n^0|n^0\rangle.$$

Suppose that $\mathcal{H}_1$ depends on time and the matrix elements $\langle m^0|\mathcal{H}_1|n^0\rangle$ are small compared to E_n^0. We can write the system state vector in the form

$$|\psi(t)\rangle = \sum_n d_n(t)\, e^{-i\frac{E_n^0}{\hbar}t}\,|n^0\rangle. \tag{A.81}$$

Having called the probability of finding the system in the state vector $|f^0\rangle$ as $P_{i\to f}$, provided that, at time $t = 0$, it is in state $|i^{(0)}\rangle$, at first perturbative order, it results that

$$P_{i\to f}(t) = |d_f(t)|^2 = \left|-\frac{i}{\hbar}\int_0^t d\tau\, \langle f^0|\mathcal{H}_1(\tau)|i^0\rangle\, e^{i\omega_{fi}\tau}\right|^2, \tag{A.82}$$

where $\omega_{fi} = \frac{E_f^0 - E_i^0}{\hbar}$ and $f \neq i$.

A.13   Born Approximation

Having called the wave vectors of the incident and deflected particle as $\mathbf{k}$ and $\mathbf{k}'$, respectively, the Born approximation to the scattering amplitude for the potential $V(\mathbf{r})$ is given by

$$f_B(\mathbf{k}, \mathbf{k}') = -\frac{m}{2\pi\hbar^2}\int d\mathbf{r}\, e^{-i\mathbf{k}'\cdot\mathbf{r}}\, V(\mathbf{r})\, e^{i\mathbf{k}\cdot\mathbf{r}}, \tag{A.83}$$

where m is the system reduced mass.

If the potential is central, the expression simplifies:

$$f_B(q) = -\frac{2m}{\hbar^2 q}\int_0^\infty dr\, \sin(qr)\, V(r)\, r, \tag{A.84}$$

where, since we consider elastic scattering, $q = |\mathbf{k} - \mathbf{k}'| = 2k\sin\frac{\theta}{2}$ and θ is the angle of deflection.

A.14 WKB Approximation

We consider a one-dimensional system of a particle with mass m and energy E subject to a potential $V(x)$ and define $p(x)$ as the classical momentum:

$$p(x) = \sqrt{E - V(x)}.$$

If the energy E is less than the potential $V(x)$ for each point outside of a certain range $[a, b]$, the eigenvalue E belongs to the discrete spectrum. In the WKB approximation, the eigenvalues are given by the relationship

$$\frac{1}{\hbar} \int_b^a dx\, p(x) = (n + \frac{1}{2})\pi \quad \text{with } n = 0, 1, 2, \ldots . \tag{A.85}$$

Equivalently, if we consider an entire classical oscillation between the two classical turning points a and b and back to a, this relation can be rewritten in the form of the Bohr-Sommerfeld quantization rule:

$$\oint dx\, p(x) = \int_D dx\, dp = 2\pi\hbar(n + \frac{1}{2}), \quad \text{with } n = 0, 1, 2, \ldots \tag{A.86}$$

where, in the first expression, the integral is extended to the complete classical trajectory and, in the second one, to the D domain delimited by it.

If the energy E is greater than the potential $V(x)$ for each point outside of the interval $[a, b]$, the eigenvalue E belongs to the continuous spectrum and we are in the presence of a potential barrier. The probability of crossing the barrier in WKB approximation is given by

$$T = e^{-\frac{2}{\hbar} \int_a^b dy\, |p(y)|}. \tag{A.87}$$

References

1. Abramowitz, M., Stegun, I.: Handbook of Mathematical Functions with Formulas, Graphs, and Mathematical Tables. Dover New York (1972)
2. Flügge, S.: Practical Quantum Mechanics, volume I e II. Springer, Berlin (1971)
3. Gasiorowicz, S.: Quantum Physics. Wiley, New York (2003)
4. Gol'dman, I.I., Krivchenkov, V.D., Kogan, V.I., Galitskiy, V.M.: Selected Problems in Quantum Mechanics. Infosearch London (1960)
5. Gol'dman, I.I., Krivchenkov, V.D.: Problems in Quantum Mechanics. Pergamon Press London (1961)
6. Gottfried, K., Yan, T.-M.: Quantum Mechanics: Fundamentals, ii edn. Springer, Berlin (2004)
7. Kogan, V.I., Galitskiy, V.M.: Problems in Quantum Mechanics. Prentice-Hall London (1963)
8. Landau, L., Lifchitz, E.: Physique Theorique vol. III (Mecanique Quantique). Mir Moscou (1966)
9. Lim, Y.-K. (ed.): Problems and Solutions on Quantum Mechanics. World Scientific (1999)
10. Merzbacher, E.: Quantum Mechanics. Wiley, New York (1970)
11. Messiah, A.: Mecanique Quantique, volume I and II. Dunod Paris (1962)
12. Nardulli, G.: Meccanica Quantistica, volume I and II. Franco Angeli Milano (2001)
13. Passatore, G.: Problemi di meccanica quantistica elementare. Franco Angeli Milano, ii edition (1981)
14. Sakurai, J.J., Napolitano, J.: Modern Quantum Mechanics, ii edn. Addison-Wesley (2010)
15. Shankar, R.: Principles of Quantum Mechanics, 2nd edn. Plenum Press, New York (1994)
16. Ter Haar, D.: Selected problems in Quantum Mechanics. Infosearch Ltd., London (1964)

© Springer Nature Switzerland AG 2019   253
L. Angelini, *Solved Problems in Quantum Mechanics*, UNITEXT for Physics,
https://doi.org/10.1007/978-3-030-18404-9

Printed in the United States
By Bookmasters